MANAGING THE PROCESS,
THE PEOPLE, AND YOURSELF

Also available from Quality Press

Integrated Process Management: A Quality Model
Roger Slater

Ethics in Quality
August B. Mundel

Quality Management Benchmark Assessment
J. P. Russell

A Guide to Graphical Problem-Solving Process
John W. Moran, Richard P. Talbot, and Russell M. Benson

Benchmarking: The Search for Industry Best Practices that Lead to Superior Performance
Robert C. Camp

QFD: A Practitioner's Approach
James L. Bossert

To request a complimentary catalog of publications,
call 800-248-1946.

MANAGING THE PROCESS, THE PEOPLE, AND YOURSELF: A PRIMER FOR OPERATIONS MANAGEMENT

Joseph G. Werner

ASQC Quality Press
Milwaukee, Wisconsin

Library of Congress Cataloging-in-Publication Data

Werner, Joseph G.
    Managing the process, the people, and yourself: a primer for operations management / Joseph G. Werner.
       p. cm.
    Includes index.
    ISBN 0-87389-193-7
    1. Production management—Quality control. I. Title.
TS155.W457   1992
658.5 62—dc20   92-25518
                                                                                                                   CIP

© 1993 by ASQC
All rights reserved. No part of this book may be reproduced in any form or by any means, electronic, mechanical, photocopying, recording, or otherwise, without the prior written permission of the publisher.

10 9 8 7 6 5 4 3 2

ISBN 0-87389-193-7

Acquisitions Editor: Jeanine L. Lau
Production Editor: Mary Beth Nilles
Marketing Administrator: Susan Westergard
Set in Garamond and Univers by Montgomery Media, Inc.
Cover design by Barbara Adams.
Printed and bound by BookCrafters.

For a free copy of the ASQC Quality Press Publications Catalog, including ASQC membership information, call 800-248-1946.

Printed in the United States of America

  Printed on acid-free recycled paper

  ASQC
Quality Press
611 East Wisconsin Avenue
Milwaukee, Wisconsin 53202

To the memory of my maternal grandparents, Luciano and Ida Meneghini, whose work ethic and perserverance remain an inspiration to me.

ACKNOWLEDGMENTS

I have learned from and worked with many talented people. The following people helped me develop my operations philosophy by challenging me to look deeper and/or showing me by their example. I appreciate their patience and thank them for their help.

They are Sister Marilyn, Thomas Wold, George Kelling, Herbert Thelen, Charles Beck, George Maddock, Richard Gamble, Harold Reese, Edward Lewis, Dennis Kozlowski, Terry Schmoldt, Roger Batten, Doris Zelinsky, Vernon Roy, Georgia Wright, and Rob Heberling.

I would especially like to thank John Adams, Robert Hoyler, James Sinnott, and my parents for believing in me.

Finally, I appreciate the help from my wife, Gretel Hartley, in this effort.

CONTENTS

Introduction 1

Part 1: Knowing the Process 7
Chapter 1: Methods 13
    What Is Quality? 13
    Quality Standards 17
    Purchasing and Forecasts 22
    The Japanese 29

Chapter 2: Environment 31
    Preventive Maintenance 31
    Maintenance Parts 35
    Environmental Conditions 38

Chapter 3: Measurement 43
    Accounting for the Process 43
    Computers and Statistics 48

Part 2: Knowing the People 53
Chapter 4: The Soft Issues 57
    Quality and the People 57
    Communicating with the Process Employees 63
    Union vs. Nonunion 69

Chapter 4 continued
    Hiring and Promoting 72
    Terminating 77
    Meetings 79
    Safety 82

Chapter 5: The Departmental Processes 87
    The Line Supervisor 87
    The Quality Control Department 94
    The Sanitation Program 99
    The Front Office 103
    The Receiving and Shipping Departments 106
    Product Development 109

Part 3: Knowing Yourself 117
Chapter 6: You 121
    Why Are You There? 121
    Managing and Improving Yourself 124
    Communicating with Your Boss 130

Chapter 7: You in the Process 137
    Style 137
    Culture 140
    Managing Change 143

Index 149

INTRODUCTION

There are thousands of process environments in this country and around the world. The sizes range from five people in a 500-square-foot building to five thousand people in a facility that covers 500,000 square feet. These process units manufacture and/or control the products and services everyone uses each day.

The demands of one process are the supplies of another process. As our society grows, the demand for additional processes increases to meet the needs of the people in it. These needs are increasingly met by processes linked together on an international level.

What is a process? In the most finite sense, just about every task a person or group of people perform in an attempt to achieve a goal can be regarded as a process. Identifying and understanding the individual tasks of a process are essential to understanding and improving the entire process and achieving the goal of making a quality product or providing a quality service. The science of the process is the interaction of the raw materials and equipment with the sociological and psychological elements of the people managing the process.

People recognize the larger processes and their results because they are easily identified. We recognize the hospital as a process environment that provides health care to the community. We recognize the automobile plant as the process environment that manufactures the car we are driving. We recognize

MANAGING THE PROCESS, THE PEOPLE, AND YOURSELF

the bakery as the process environment that manufactures the bread used to make our sandwiches.

However, most people recognize, on a minimal level only, the costs, the planning, the experimentation, and the different processing procedures that were required to produce each of the components in the surgical room, the automobile engine, and the wheat in the bread. We acknowledge the complexity of these processes at a level of consciousness that holds them far enough away from our personal trials and tribulations to allow each of us to perform our role in our process but close enough to gain a level of confidence that this product or process will be safe and beneficial. We put our faith in the other people working in the other processes. As one commercial stated recently, "I want this car to be safe, because someone's grandmother might be driving it."

The belief that each part of the process is managed in a safe and ethical manner is really beyond the normal level of understanding. For example, when considering our bread, we assume that the chemical company that supplied the fertilizer to the farmer for the wheat he or she was growing was adequately tested. We assume the fertilizer was found to be safe for the farmer, the wheat, the ground, the ground water, and next year's crop of soybeans. We assume that the farmer applied the fertilizer in the prescribed manner. We assume the grain elevator and flour mill stored and processed the wheat to avoid contamination by rodents and equipment. We assume the bakery mixed the wheat with ingredients manufactured in other controlled processes. We assume that all of these processes were in control and that the peanut butter and jelly on the bread that is being eaten by our children is also safe. If we did not believe that every process producing the products we use was controlled, we would be paralyzed by fear from using any of them. If we had the knowledge of and seriously considered the number of variables that can go wrong in a process, our rational capabilities would be overwhelmed.

This implicit faith in each other's process efforts is the greatest and most important characteristic in the success of humans as social creatures. When people live together, they have to help each other. The beauty and strength of this faith is

INTRODUCTION

supported by the fact that almost all of the products and services provided by the different processes around the world are almost always safe and efficient.

The whole situation is really a sociological paradox, because nobody has time to grow their food, sew their clothes, make their shoes, teach their children, build their house and car, and still have time to watch the Cubs win the World Series on their homemade television. We generally don't have time to pick out vegetables in a farmers' market, which is about as close as most of us come to getting back to the basics. We don't even have time to find the farmers' market! We really do need each other to help perform those tasks we cannot perform in the countless processes on which we rely to live.

Yet, how different are the underlying processes that govern each of these processes? How different are the people who work in each of these facilities? How different are the owner's objectives for the processes and the products?

There are fundamental similarities that control the success of each of the processes, regardless of the type of process, the location of the facility, or the people associated with it. Every task a company performs is part of the process and, therefore, is governed by process control methods. Understanding these common points can be beneficial to the process and the people working in the process. Understanding the process will make the people working in the process happier with their jobs, because knowledge is the most important ingredient in success. The bottom line is that everyone and every company wants to be successful in their own way.

This book is divided into the three sections that govern all processes. Each section contains discussions of the different subprocesses in that area. The three main sections are:

1 *Knowing the Process:* understanding the science of the process in the operation and determining how the main goal of the process is affected by the subprocesses in the operation and the processes outside the operation;

2 *Knowing the People:* recognizing people as individuals, developing their strengths over their weaknesses,

MANAGING THE PROCESS, THE PEOPLE, AND YOURSELF

and understanding how each person should perform his or her role in the process;

3 *Knowing Yourself:* knowing your strengths and weaknesses and how they need to fit into the process.

The three elements interact at each stage of the process on each process variable. Analyzing, managing, and explaining process issues to the people in a productive manner can only be accomplished when the effects of all three elements are considered. Improvements in the process are a result of understanding how the process, people, and "self" issues interact over time.

Developing and expanding this understanding is a prerequisite for continual improvement. Developing and expanding the knowledge of the interaction between the three variables is the essence of the "Managing the Process, the People, and Yourself" (MPPY) philosophy.

What will it take to be a successful process manager in the next decade and beyond? The successful process manager must be willing to do whatever it takes to get the job done. He or she must be willing to expand the current level of knowledge of the science of the process and to devote himself or herself to the challenge of continually improving the process. He or she must be willing to ask "Why?" and "Why not?" while striving to understand the process. He or she must be able to communicate with, motivate, and teach the people about the science of the process and to bring an enthusiasm that makes learning about the process interesting and challenging for the people.

The best managers in process environments will have the skills of teachers or educational psychologists. Engineering and financial skills, which have been emphasized in the past, will become skills that support the teaching skills. The best managers will develop a continual improvement philosophy by understanding and managing the group dynamics of their process.

Finally, each manager will need to be his or her own inspirational leader, because the corporate controls over their

INTRODUCTION

decision-making processes will be decreased. Responsibility will be pushed farther down into the organization as layers of management are eliminated. These managers must be able to maintain a clear and consistent picture of where the process is going as it moves through the "hills and valleys" on the road of continual improvement.

At different points in the book, important MPPY rules or guidelines that help define the focus of the new process manager are highlighted. Some of them are shown below to give you an idea of what to look for in this book.

MPPY Guidelines

1. You can reduce the cost of your product without affecting the quality of the product. You can improve the quality of your product without increasing the cost of the product.
2. Identify what works well and stay with it.
3. Your process talks to you every second of every day.
4. Work is a social situation.
5. All parts of the process are managed by people.
6. Develop the strengths of the people to overcome their weaknesses.
7. Control the process. Do not let the process control you.
8. Any activity that is not adding value to the process should be eliminated.
9. A process will go "downhill" faster than it can go "uphill."
10. You must HAVE it to be successful: Honest, Accessible, Visible, and Enthusiastic.

PART 1
KNOWING THE PROCESS

Can you describe from start to finish how raw materials, soon-to-be products, move through your process area? Can you describe the effect variations in the different process control parameters have on the quality and efficiency of the process? Do you understand how the equipment affects the process and how the process affects the equipment? Does your process have a system to measure its performance to the performance targets? Are you controlling the process or is the process controlling you?

No matter what kind of process you are working with, certain parameters exist that help all processes run more efficiently. Understanding the science of the process implies understanding these parameters. Understanding the individual parameters and their effects as a group on the entire process is the only way to achieve a level of quality output that makes the process and the people involved in the process a success.

Individuals working at specific tasks on the line must have and understand the specific process parameters for their subprocess. The quality of the performance of the subprocess activities will affect subsequent processes applied to the product. If one part of the process goes out of control, all parts before and after the derelict process may go out of control.

It is obvious how the processes after an out of control process will go out of control, but how can it affect processes that preceded it? Well, it's easy to fix something that's broken if you know what caused it to break, but many times you aren't sure what is broken. The problem can be small or cumulative and people start changing different parts of the process to correct an "error"—in a process parameter that was perfectly fine.

For example, I was working in a process that comprised 14 subprocesses. We could not get the product collated and packaged efficiently, which were subprocesses 11 and 12, respectively. We spent a considerable amount of time analyzing and modifying the equipment at these stages of the process without success. We then suspected it was a problem with the first subprocess and made some changes there also. These changes were not successful. When we finally determined that the problem was in the ninth subprocess and corrected it, the packaging operation improved within minutes!

Without a fundamental understanding of the *entire process*, you can never be sure of the impact a change in one part of the process will have on subsequent processes applied to the product. Without a complete understanding of the process, it will take you much longer to identify which process is out of control so you "don't fix what isn't broken."

In the 1980s, many companies reduced costs by reducing headcounts. Most of these cost-saving opportunities have had little, if any, real effect on the ultimate goal of the business units, which is to improve the knowledge of and thereby the efficiency and quality of the main process.

The companies that will survive in the 1990s and beyond will generate permanent and significant savings by digging deep into the technical aspects of their processes. Understanding the science of the process will allow the company to identify the cost elements that have a direct impact on product quality and those that do not affect quality.

These companies will focus the efforts of and reward the process managers working to improve the understanding of the science of the process.

The best process managers will make it their goal to spread this knowledge to all of the people involved in the process. Information about the process will be pushed up, down, sideways, and around the communication ladder in the process. When information on the process is pushed, reviewed, and discussed at all levels of the process, people will be able to make the right decisions to improve the process.

Push information and knowledge to get the right decisions. You cannot push decision making on the people without first understanding and establishing the technical parameters of the process. Pushing decision making without the knowledge of the science of the process will reduce the efficiency of the process and destroy the enthusiasm of the people. People cannot be held accountable for the efficiency of their process or the quality of the product if they and the management group do not understand how their process is supposed to perform. They must first be held accountable for determining how the process is supposed to work; once they understand this principle, efficiency and quality wiil follow.

The best companies will monitor the progress each unit makes to accumulate this knowledge. The best companies will incorporate the new ideas and procedures into their other processes. These companies will succeed by standardizing good process ideas quickly and eliminating or changing processes that no longer add value to the product.

CHAPTER 1
METHODS

WHAT IS QUALITY?

What does quality mean for your product? Quality is applied as a general performance term to all products, but the quality goal of your product, set by your company and not you, determines how you should manage your process. The definition of quality does not vary between products, but the application of the definition to the process and product does vary. Consistently meeting and exceeding the quality goals set by your company means you are making a quality product.

For example, an automobile, car F, has a particular quality standard. Another automobile, car C, also has a quality standard. Each part of the process used to assemble these vehicles has specific quality parameters. Each product is manufactured according to these quality specifications. At the end of the process, each product is pronounced to be a quality product. The consumers trying to decide which car to purchase examine all of the differences and the costs of the differences. The decision to buy one of the vehicles is based on the value they assign to the differences relative to what they can afford to buy.

The ultimate success of any process is measured by the demand for its product. The demand for a product is a result of the value consumers apply to the product. The value of a

PART ONE: KNOWING THE PROCESS

product is a function of the cost and the quality. The sales and marketing processes of a company monitor the consumers' value perceptions to determine how to sell the product relative to the value perceptions of comparable products.

If the product quality meets the desired specifications of the consumers for the price they can pay, then consumers may use the product. They may also choose not to use the product, because many factors contribute to the value perception of your product in the purchasing decision. Some factors, such as price, quantity used, how and how often it will be used, and average life expectancy of the product, are tangible. Other factors, such as the aesthetic value, the image associated with owning the product, the packaging, the color, the service and the distribution programs, are less tangible.

Your process may be meeting and exceeding the quality goals set by your company, but the value of the product to the consumer, relative to all of the tangible and intangible factors in the cost-quality relationship, is less than the value required to purchase the product. In this situation, the company has set the quality goal either too high or too low.

In a free market or a controlled market, consumers will always be able to determine the correct value of a product to them. The best example of the effect of a low-quality goal on consumer purchases can be found in the U.S. automobile industry during the last 10 years. An example of a quality goal that may be too high relative to its perceived value is in nationally distributed "fresh" processed food products.

Some companies may be open to your suggestions and ideas about the tangible and intangible factors. You will find that your ideas will have more acceptance if your process is making quality products in an efficient manner. So, make sure you have your process in control and understand thoroughly how the different subprocesses affect the quality of the product. This knowledge can be used to determine the future of the product.

For example, how can the process be changed to develop new products for the sales and marketing groups? Many times, a new product can be developed from the same equipment with only minor modifications. A less complicated change

METHODS

might be as simple as changing the case size or reconfiguring the packaging pattern so the pallet fits into the customer's warehouse more efficiently. The people working with these parts of the process every day can be a very good source for these kinds of suggestions. Make sure you have a mechanism to get feedback from the customer to all levels of the process.

A less-than-expected sales volume may be the result of one or more of the intangible value factors. Correcting this kind of problem may be easier than identifying it. In these situations, the decrease in sales may be blamed on the quality of the product. If the quality of the product is lower than expected, then you have some work to do. If you understand your process and have the data on your performance to the quality specifications, then you will be in a better position if the product quality is unfairly blamed.

There will undoubtedly be periods of frustration with how your product is handled by the company. Here is a "Managing the Process, the People, and Yourself" (MPPY) control rule: *Operations people will never understand marketing.* But don't worry, the marketing people understand it and operations, too!

My favorite line on marketing and advertising is attributed to Alfred Bloomingdale, the founder of Bloomingdale's. He said, "I am sure 50 percent of our advertising dollars are worthless. I am just not sure which 50 percent." He said this in 1925. If you have had the opportunity to sit in on any marketing meetings about your product, then you must be wondering what state the world would be in today if operations had progressed as far as marketing since 1925.

Despite all of the demographic information on consumer preferences and behaviors, it is difficult to predict which products will succeed and which will fail. The large number of new products, "me-too" products, and brand extensions entering the market every year make it more difficult for a company to identify why a product will be successful and for the consumer to determine where the value of the product is on the cost-quality continuum. The tangible competitive differences become so small that they become intangible. This phenomenon may explain why brand loyalty has become harder and more expensive to define and maintain.

PART ONE: KNOWING THE PROCESS

There are processes within the mind of the consumer that cannot be catalogued. These processes determine the value of the cost-quality relationship in all purchasing decisions. These processes vary from person to person, region to region, and country to country. These differences make it more difficult to understand the perceived value of a product and to predict how it will perform. Attempting to predict the degree of success a product will achieve is important for determining the expected financial return on the investment in the process.

You can control many factors in the cost-quality relationship of your product. You can control the quality and rate of consumption of the raw materials, the labor configurations that run the equipment, the program to take care of the equipment, and the culture within the process environment.

You should be armed with suggestions on how to improve the factors you control and the costs of these improvements. You must know your process well enough to know where to improve it. An MPPY operations rule is as follows: *You can reduce the cost of your product without decreasing the quality of your product.* The converse is also true: You can improve the quality of your product without increasing the cost of the product.

Reducing or changing the product quality is the easiest way to cut costs. It happens all the time, because it is the quickest way to save money. It is usually the only way to reduce costs for those people who do not understand the science of their process. This method may provide short-term relief, but it most certainly will decrease the chance of success in the long term. You cannot cut costs, any kind of cost, unless you understand the exact role it plays in your process and how it affects the quality of your product.

Once you know your process intimately, the "doors and windows of the house" of cost reduction and quality improvement will open for you. Realizing quality improvements without increasing costs is a sign that you are definitely beginning to understand and control the science of the process.

You should always strive to reduce your production costs while maintaining or improving product quality. Lower costs translate into better prices. Better prices at the same level of

METHODS

quality increase the value of the product in the cost-quality equation. Lower costs give your sales and marketing people more room to promote the product and put pressure on the competition. It may not guarantee the survival of your product in the marketplace or even your facility, but without a strong desire to reduce costs, your process and product will eventually lose to the competition. Someone or some company somewhere is always looking for ways to make a better "wheel." If you give them the opportunity, it will be exploited.

QUALITY STANDARDS

When a problem arises in the process area, people spend a great deal of energy correcting it. People who normally stay in their offices may come into the process to find out what happened. But does the extra help add value to the problem solving analysis? Was the analysis managed in the correct manner? That is, was a thorough investigation of the cause and effect conducted? Were all of the people involved in that part of the process included in the discussion of the problem and its resolution? If a piece of equipment failed, was it noted on the work history of the machine? Did everyone react in the proper manner to minimize the loss of product? Should any of the product be placed on hold?

Whether a problem is reccurring or new, it can provide great insights into the workings of the process when analyzed properly. Never miss the opportunity to learn everything you can from each problem that arises in the process. This analysis is the only way to prevent the problem from recurring. In some cases, a simple checklist can be created to help all of the people remember what to do and what to look for. Checklists are effective, because process people have a lot of information to remember and everyone has different memory skills.

Is it unusual to spend a lot of time correcting a problem? No. But it is unusual that the same amount of time and energy is not devoted to analyzing and thinking about the process when it is *running smoothly!* Managers tend to stay in their

PART ONE: KNOWING THE PROCESS

offices and machine operators tend to relax when the process is running smoothly. The best time to go out into the process environment and think about what is going on is when it is running smoothly, because there is no specific problem on which to focus. Some of the best insights into the science of a process can be achieved by identifying the conditions that exist when it is running smoothly. It is the best time to see how the process is meeting the desired quality standards.

When a process starts to get out of control, it is fairly easy to figure what changed if you have identified the conditions before the change. A simple MPPY operations rule is as follows: *Identify what works well and stay with it.* This simple concept is missed by many process managers. It works for coaches, teachers, parents, athletes, and the sales and marketing programs of a company.

Knowing what works well and staying with it does not mean you should never change the process. Knowing what works well should become the basis for improving your quality. It will provide you with insights into the range within which each process parameter functions in each of the subprocesses. Understanding your process means understanding the limits of these ranges. Improving the quality of your process will come from reducing the range of tolerance for each of these process parameters. You can identify these ranges by using statistical process control techniques.

Remember that very few process parameter ranges are uniquely independent of the product quality and the efficiency of the line. These parameters almost always overlap with ranges of other parameters behind, adjacent to, and ahead of them in the process. Changing one parameter will affect other parameters and their control ranges.

While the challenge of finding out how all of the ranges interact to affect the quality standard can be quite difficult, there will usually be two or three "killer" parameters that affect most of the other parameters. The "killer" parameters can be easily identified by concentrating the efforts of the process staff on the "Why is the process working well today?" analysis. The interaction of the parameters can be measured using analysis of variance models.

METHODS

Another area of quality standards that must be given attention is the quality and consistency of the materials consumed by the process. These materials have specifications the supplier must meet. These specifications should be based on the knowledge of how the materials interact with the process. Someone in the facility should record and monitor the changes in these materials as they are received. The changes in the raw materials should be related to changes in the performance of the process.

Keep visual samples if you have the room. Store the records of the tests in a database for easy comparison to the historical performance of the process and quality data.

Don't assume that the results of a test performed by the supplier are always correct. Use an outside laboratory if you do not have the resources to test the materials. Try to get a quick turnaround on the results of the test. Ideally, you would like to have the data before you use the material, but trying to keep inventories at a minimum will make this very difficult. Minimal back-up inventory of raw materials is a very good reason to ensure that what you just received will work.

If your analysis reveals that a material specification may be reducing product quality or the performance of the process, call the vendor. Ask the vendor about the product. Discuss how the product is being used in the process.

Call other suppliers of the same product and discuss the problem you are having with the material. You may want to have a sample of the material from another supplier to test. This can help determine whether the problem is related to the material, the equipment, or the process.

Get your corporate purchasing group involved early on in the discussions with the supplier. If you need to request a change in the material specification, it will be easier if the purchasing group has been informed. If the change is going to increase costs, make sure you have the records on the performance and quality of the product to back up your request. A company committed to improving the quality of its products will support your request if you can demonstrate the difference it will make.

Companies committed to getting bonuses for favorable purchasing variances will be reluctant to make a change that

PART ONE: KNOWING THE PROCESS

increases costs. Remember, most of the corporate purchasing gurus have never worked in a process environment. It will be difficult for them to see the relationship between the specification issues and the quality of the product. You might suggest changing the specification on one delivery to conduct a longer test of the new material. Have your facts ready and make sure the product really does improve the process.

I had an experience with an ingredient that was causing significant problems in the process. The ingredient varied greatly from shipment to shipment, and we were constantly readjusting the process to compensate for the variations.

I found another supplier of the ingredient that more closely matched the needs of our process and also was more consistent in quality between loads. The new supplier was a little more expensive, and corporate purchasing was not sure it was worth it. We demonstrated that it was better for the process by running test loads of the new product.

This comparison test was not good enough. We then bought the same test equipment for our lab that the original supplier was using, because, according to their quality personnel, we were not testing the material correctly. We really didn't understand our process. The new test equipment still showed variations in the product.

The relationship with the original supplier ended at a meeting attended by the general manager of the supplier, my boss (the plant manager), our vice president of corporate purchasing, and me. After a long, increasingly heated discussion about the quality variations, the general manager said that there was nothing wrong with his product or process and that was what we would get.

We changed suppliers shortly thereafter. The supplier lost the account, not only because the product was not meeting our standards but also because he refused to work with us anymore. He would not listen to or address the needs of his customer. It is possible that the supplier's process was not capable of meeting our needs. But, because they were the largest supplier of that particular product in the area, they could not imagine how there could be a problem with their process. They lost an account and the opportunity to learn more about the limits of their process.

METHODS

An instance may arise in which the materials that are consumed by the process are fine, but some of the materials that *support* the process are causing problems. Make sure you have the correct cleaning solutions and cleaning equipment. Make sure the maintenance group has not recently changed the type of lubrications used in the process. Sometimes little, subtle changes in the support materials can cause big changes in your ability to control the process.

Gather information about your process from all sources. I believe you can find out almost anything you need to know in five or fewer phone calls. Attend trade shows. Visit other plants. Talk to equipment manufacturers. Always share this information with the people in the process. Discuss how the new knowledge can improve the quality standards. Discuss the specific performance feedback the material and equipment should be giving to the people.

What happens when a product quality standard is not met and this is not discovered until after the product leaves the facility? In cases in which harm to the company or consumer will occur, the company will issue a recall. The decision to bring the product back protects the consumers and the company. If you are aware that your process has gone out of control and the desired quality standards are not being met, the worst thing you can do is nothing.

Take immediate action by placing the product on hold. If necessary, stop the process. Make sure your records account for all of the product in question. Conduct an *immediate* and thorough investigation of the incident; if you wait too long, some of the clues may disappear, change, or be forgotten. Hopefully, you have developed an open and honest atmosphere in the process, and the people will talk about what happened.

The money lost by holding, and perhaps destroying, the product will be small compared with the potential loss from allowing the product to go on the market. If your company is more worried about the costs than the risk to the consumer, I would think seriously about changing companies.

I knew of a company that held $40,000 of a food product because one plastic washer from a joint in the process pipes was missing! They were going to sort through the product and

PART ONE: KNOWING THE PROCESS

determine which of the 30 pallets should be isolated for a detailed inspection.

Many things should be remembered during a product hold or recall. The best way to make sure everything happens correctly is to have a program and checklist to deal with a recall. The checklist should have the phone numbers of the appropriate corporate and process personnel. Make sure the procedure is clear enough for the second- and third-shift personnel to execute. Remember, the worst thing you can do is nothing!

A final important point in the method to achieve the quality standards is that the people on the line are responsible for the product quality. Yet, the ability to succeed is contingent on the tools and information they have at their disposal. It is your job to make sure the line operators and the supervisors have all of the information and resources they need to meet and exceed the desired quality and line performance standards. Making the people responsible and holding them accountable for the quality of the process presumes that they understand how they can and do affect product quality.

PURCHASING AND FORECASTS

There are two types of purchasing scenarios that you will have to deal with at your facility. In the first scenario, your corporate office will do the negotiating and actual purchasing of the materials. You will release these materials from the supplier into your process. In the second scenario, you will perform all of the purchasing tasks from the price negotiations to the release of the materials. Let's look at each scenario.

If your facility is part of a large company, then a corporate purchasing department will probably be responsible for some of the materials the process consumes. There are advantages to this arrangement, because having several small contracts combined into one contract for a single item provides the corporate purchasing agent with some pricing leverage. The supplier can be motivated to make a better deal if the volume is greater, because the supplier can have longer runs of the item in its process.

METHODS

However, just because you have a corporate purchasing agent doesn't mean he or she will save the company money. It will still depend on the negotiating skills of the agents involved in the deal. I knew of an instance in which a subsidiary obtained a significantly lower price than the parent company, which had a much larger volume requirement, through better negotiating.

You can improve your negotiating skills with seminars and practice. Always ask for a better deal; you have nothing to lose if they say no. Keep them waiting. If they want the business badly enough, they will work with you.

The most important factor for your process in the corporate purchasing arrangement is quality. Don't accept a low-quality product from a supplier just because the purchasing agent saved money.

In the 1990s and beyond, companies and suppliers will come together to work out price, quality, and delivery issues. The supplier hopes to gain a long-term advantage in the relationship and is motivated to work with the company's program if it wants the business. This arrangement provides the company with some of the benefits of vertical integration without the headaches and investment.

If your purchasing process is not centralized at corporate and you are responsible for it at your facility, then assign someone to it. If your job is to keep the process going, then you won't be able to spend very much time on purchasing materials. Purchasing the right materials at the right price takes time.

The purchasing agent, and the operators and supervisors, should review the process control chart records on how the supplier's product performed in the process. This performance feedback should be shared with the supplier to help them make the right product for your process.

The agent should develop long-term relationships with these suppliers. Keeping the quality of the raw materials consumed by the process consistent will allow you to fine tune the process. A good, long-term supplier will also do a little extra for you if volume increases unexpectedly and you need more materials on short notice.

PART ONE: KNOWING THE PROCESS

Your purchasing agent will make purchasing decisions based on the actual usage of the material in the process environment. It is not an easy job, and it is usually thankless. If everything goes well, nobody notices; but if some material runs out, everyone gets excited.

In most instances in which the process unexpectedly runs out of a raw material, the feedback system on actual usage in the process broke down. When you do run out of a material unexpectedly, treat it the same as you would a processing issue. Find out why and how it happened. Recreate the "beginning inventory plus purchases minus usage equals ending inventory" formula for the item.

Verify the count on the receiving ticket. Perform live audits of the actual usage of the item in the process and adjust the standard, if necessary.

Many of the new computerized Material Requirements Planning (MRP) systems have standard usage and scrap figures built into their calculations. Maybe the standard figures no longer reflect the actual process activity.

If you are still running out of materials, then have a short scheduling meeting with your supervisors, key operators, and purchasing people each week to review what is happening next week. Everyone should be checking the materials daily, until you get confidence in the weekly analysis.

The purchasing questions of when and how much to order are always the most difficult to answer. The answers to these questions can be affected by seasonal price variations in the product you are buying and/or producing for sale. The correct order quantity can also be affected by the amount you order, the lead time of the supplier, the flexibility of the processing system, and the amount of storage space available at the facility.

Purchasing agents must rely on a sales forecast for the "when and how much" answers. The responsibility to have all of the materials ready on time is shared with the sales and marketing groups through the forecast. There should be a weekly meeting with these groups to review and update the forecast and its impact on the process environment. The forecast should provide estimates of the weekly, monthly, and

METHODS

annual sales volume for each item. Special attention should be given to promotion volumes to avoid having specially marked or prepriced products left in inventory that cannot be sold when the promotion is over.

How many of us have spent time learning the economic order quantity (EOQ) formula? This formula is designed to help answer some of the questions the forecast generates. It has always amazed me how many people have been trained on this purchasing technique by doing EOQ problems in a classroom, but how I have never met one person who actually used it. The "Essentially Outdated Quantity" does not fit actual purchasing situations very well.

Why? Because forecasts are never that accurate or predictable, even when the different smoothing techniques for usage variation are used. The models cannot predict how the products will sell and how the process materials will be consumed. The forecast should be reviewed each week by the different groups due to its variability.

Consumer purchases of some items may fit some of the forecasting models, but the accuracy of any model is diminished by coupons, rebates, discounts, and competitive responses. The sales and marketing groups utilize these different programs to generate volume. The sales volume usually increases, but the standard error in the forecast also increases.

It is the uncertainty of "when and how much" will actually be sold that drives purchasing and process managers crazy. The managers will either be scrambling to find materials, people, and machine time to make the order, or they will be looking for ways to keep the people busy, because the order is too small. Make no mistake about it, though, it is always better to have orders no matter how hard everyone has to work to fill them.

Every day, more and more companies are getting actual sales and order information sooner. In some cases, these orders become part of the process system within 24 hours. The clothing retailers and snack manufacturers are examples of companies that use these quick-turnaround information systems. I read recently about a company that had an order in its system that was processed in 2 hours!

PART ONE: KNOWING THE PROCESS

The nice aspect of these systems is that they respond to changes in customer demand almost immediately. This instant information system will generate more sales, because it is reacting to changes in the purchasing patterns of the consumers as they are occurring.

Yet, balancing the inventory, the service level, and changeover issues in a quick-turnaround process environment becomes more difficult. Raw materials and finished goods inventory are required to service the needs of the business. Determining the minimum level of inventory will be a function of the desired service level against the forecasted orders, the length of time to convert the raw materials into a finished unit, and the cost of carrying the inventory to meet the service level. The longer it takes to make a unit, the more money the company will have invested in the inventory.

Facilities operating in environments that must have a quick-turnaround order cycle should be looking at those parts of the process that can be standardized across the different product lines. For example, maybe one case can be used for three different items. Maybe the same size carton can be used for several products to avoid changeovers and downtime on the equipment. This arrangement provides the purchasing agent with more flexibility to meet the consumption needs of the line.

Marketing decisions on new products and packaging materials have to consider the capabilities of the process equipment for the same reasons. I had an experience in which a company's packaging equipment was dedicated to one style of packaging. Marketing wanted an entirely different packaging concept with the same product to create the illusion of a new product, repositioned for the perceived needs of the current market. It was going to cost several million dollars to modify just one of the process lines to accommodate the new idea. With better long-term strategic thinking, the line could have been laid out to handle some of the new ideas more efficiently.

An alternative approach is to design flexible manufacturing systems that can be changed over quickly to produce a different product or line of products. This option is more expensive and difficult to design and justify. A flexible system that attempts to incorporate all of the possible product configurations, current and future, into a single process line would be an engineering nightmare.

It is also very difficult, if not impossible, to design a process that is completely flexible to changes in the raw materials for the different products and that can still operate at profitable speeds and performance levels. The maximum amount of flexibility any process can achieve is related to the degree of redundancy in the materials consumed and the manner in which they are consumed. The maximum amount of flexibility required by a process line is contingent on the expected sales volume. With larger sales volumes, dedicated process lines are the most efficient and profitable, because processes and people thrive on consistency in the daily functions.

So, how does all of this affect the purchasing agent? All of the companies in the marketplace are moving toward serving the customers more quickly and efficiently. With or without an accurate forecast and the appropriate manufacturing system, a greater burden is placed on the ability of the purchasing agent to forecast the needs of the plant.

Unfortunately, the best plan some purchasing agents have is the one in his or her head. This plan is based on past experience, which usually leads to more inventory of some items than is necessary and not enough inventory of other items. This type of planning is called "just in case." Sometimes an incorrect "just in case" judgment turns into a "just in time" delivery, because there is too much process data to remember and interpret correctly. A "just in case" planning scenario costs the company money, because it is not turning material expenses into sales income.

It also takes space to store materials, and most plants do not have much space. It is too expensive for the company to keep open areas out of production. Generally, purchasing and holding large amounts of a raw material is not practical unless the price varies significantly over time. The purchasing agent may be able to buy a contract on the material to hedge against price fluctuations. Potato chip manufacturers will actually purchase and hold potatoes in storage from the fall harvest until the first summer harvest to control price and quality variations.

PART ONE: KNOWING THE PROCESS

When a facility has a limited amount of storage space, a good supplier close to your plant can be invaluable. Maybe the supplier has space and can inventory some of the items for you. Maybe you can use the material right from the trailer that delivered it. Some companies unload incoming materials from the trailer and replace them immediately with the finished goods.

Another option for the small facilities would be to find a warehouse nearby that will hold and deliver a mix of the items you require that day. This option will add some costs to the process for double handling, but it may get you through a busy period until you can determine whether a building expansion is required.

It is important to realize that all of these issues affect every purchasing agent and process environment producing or handling products consumed before, during, and after your process. Some of the suppliers you are doing business with today may not be the same ones you will be doing business with in 10 years. They may not have the desire or resources to meet the new demands placed on them. Make sure you work with your purchasing agent to push the supplier for help. Be fair, but don't let loyalty stand in the way of your own survival if a supplier cannot keep pace with your needs.

Unfortunately, the frustrations of the process purchasing agents will increase in the 1990s and beyond. Companies squeezed for profits and pressured for growth have made and will continue to make commitments to service customers more quickly and efficiently. With some exceptions, most of the purchasing agents will be forced to play "catch up" until an information base with the appropriate level of detail is developed. The accuracy of the forecast and the ability of the process environment to meet the desired service goals of the company are inextricably linked.

METHODS

THE JAPANESE

A considerable amount of discussion and analysis has been focused on the ability of the Japanese to establish efficient processes that can consistently produce quality products. They have achieved this position because they understand the manufacturing process and the people in the process. They have created a culture within the facility that supports the process. In addition, the culture in their country has rewarded people with strong ties to tradition, loyalty, and teamwork.

These cultural characteristics have been developed in the transplanted Japanese companies in the United States. The transplanted Japanese companies have successfully developed this culture because they invested the time and money to find the right people to work in their facilities and to train the new employees on the details of the process in the facilities.

The success of the Japanese in the United States created a "me-too" movement in the 1980s in American companies. These companies moved to adapt the Japanese style of management and culture to their processes. Production people were organized into quality circle teams. Consultants and corporate managers were brought into the facilities to tell the line operators the line belonged to them. "They really know what's going on, so let them run the process" became the management phrase of the day. Experts were brought into the facilities to instruct the line operators on the value and techniques of statistical process control.

The 1980s will be looked on as the decade during which we tried to imitate the Japanese but failed. Why didn't it work? There are a couple of reasons. The first and most important reason it didn't work for all of the companies who tried to emulate this style was that they did not understand the science of their processes. Creating teams of people armed with control limit charts and inspired by the equivalent of the half-time "let's go get it, gang" speech did very little to change or improve their knowledge of the process. They didn't understand that the Japanese brought their knowledge of the process with them.

Another reason many companies failed to see improvements after "adopting" the new culture was due to their lack of

PART ONE: KNOWING THE PROCESS

commitment to the new concept. Many managers believed that turning over the imaginary reins of control to the line employees would result in immediate improvements. If product quality didn't improve and efficiency didn't increase, there was something wrong with the teams, not with the management. The line operators recognized this shallow "Pontious Pilate" approach to the new culture, and the new culture became just another program to them.

There were companies that did have some success with the new concept. These companies either understood or were on their way to understanding how their process worked. Some of these companies learned about their process by becoming suppliers to the transplanted Japanese companies. They identified, developed, and encouraged the explorers in their operation to delve into the process with their control charts.

The successful companies realized that a process comprises many overlapping parts. It takes time, effort, and the full support of management to break down the process into these fundamental parts. There are very few quick fixes in a process. Process control must be a never-ending improvement of the science of the process program.

At the end of the 1990s, the successful products will be a result of the company and the employees working together to understand the science of the process. They will have learned how to create a culture that supports the right decision with the right information.

Achieving the optimum cost-quality relationship is a responsibility shared by every member of the organization. The Japanese do not possess any great secrets or methods in the area of process management; they just work very hard at it and were the first people to apply it effectively. More importantly, they have shown the rest of the world how effective and successful the method can be. By the year 2000, the Japanese will be able to see their methods applied in processes all over the world. As a result, the value of all products in the cost-quality relationship will increase no matter where they are produced or who produces them.

CHAPTER 2
ENVIRONMENT

PREVENTIVE MAINTENANCE

Whether your preventive maintenance program is simple or complicated, the goal of the program is to keep the line running. How do you keep the line running? It takes a day-to-day, week-to-week, month-to-month, and year-to-year effort. Some parts of the program will be reactive. It will form as information about the equipment and process becomes available. Some parts of the program will require a long-term planning approach. You have to anticipate the problems before they occur.

A good preventive maintenance program requires a knowledge of your equipment and your process. Keeping the process running efficiently is a result of the quality of the "marriage" between these two elements. Just as in an actual marriage, your knowledge of the process and the equipment will improve as you work with them. In the beginning, you are overwhelmed by the complexity of each element of the process. Then, you settle in a little but are still surprised by your partner. Later, you are comfortable and know what to expect. You can deal very effectively with the little surprises. In time, you manage the big surprises effectively and take them in stride. The pleasure you get from the growth in the "marriage" becomes deeper, and you appreciate its subtleties more each day.

PART ONE: KNOWING THE PROCESS

Whether it is your marriage or your process, you have to work hard at improving it. You have to keep communicating with each other. A sure sign the "marriage" in your process is on rocky ground and headed for "divorce" is when production is blaming the equipment and maintenance is blaming the operators. It can reach a point where even productive observations are not communicated and developed. Both groups are wrong and are contributing to the problem when the "divorce" reaches this stage in the litigation.

You communicate with your process through all of the activities you use to manage the line. An MPPY operating rule is as follows: *Your process talks to you every second of every day.* When you realize that the process is talking to you, you have 50 percent of your solution to knowing the equipment and knowing the process; the other 50 percent of the solution is learning the language it is speaking to you.

The language the process speaks is complicated and different at different points in the process and on different pieces of equipment. It takes everybody working together to assemble the lexicon of the line. The machine operators should be selected for and encouraged to develop their ability to see, hear, and feel changes in the process. Good operators can detect subtle changes and noises in the performance of the "marriage" in their area. I knew an operator who could hear if a bearing was starting to go bad on a piece of equipment in a room that was loud enough to require hearing protection!

It is essential that the supervisors and maintenance crew react to these opinions and insights. A single opinion or observation from a supervisor, maintenance person, or operator may not be exactly right, but it might be directionally correct. You should encourage the development of an open, nonjudgmental atmosphere that elicits all observations on the "marriage." Take the time to track every lead to its end, and give the feedback to every person involved in the observation. The next time a change is observed, the line operators will have one more piece of information. Keep the "idea faucet" wide open to learn the subtleties of the process!

Once your process is off the correct track, it will affect the quality of the product, the attitudes of the people, and the

ENVIRONMENT

performance of the machines. It will take even the best tracker some time to retrace the path and find the correct trail. All of the time spent looking for the right trail is time your process is not running at its optimum level. An MPPY operating rule is as follows: *If one machine is acting up, then it is probably the machine and not the process.* If all of the machines start causing problems, then it is the process that has gone out of control.

Because no two machines are exactly alike, some of the "weaker, more sensitive" machines will react more quickly to changes in the process. These machines can be used as barometers of your degree of process control, but first you have to know the equipment well enough to know which ones are the barometers.

Maybe it is an older machine in which the operating tolerances have decreased. Maybe it is a newer machine with slightly different tolerances than the original machines. As your equipment ages, you will adjust the process in small, unnoticeable increments to compensate for the deterioration of the equipment. New machines brought into the process to replace older equipment may not be able to accommodate these changes to the process.

Again, be careful about changing the machines when the process is out of control. It may be difficult to get back to the original starting point. Record the settings and time before you make a change.

It is difficult to develop and maintain a good preventive maintenance program. It is difficult because, just as in a marriage, you have to live with it and work at it every day. There are other things you can do that will contribute to your success with the preventive maintenance program in your process: Take your machines apart, take pictures, label all of the parts, and make sure your parts department is aware of and can get all of these parts.

Use these photos and videotapes as training manuals for your maintenance people and your line operators. Yes, the line operators should be trained to perform basic maintenance, too! The more information they have about the machine, the better they will be able to understand its language.

PART ONE: KNOWING THE PROCESS

Take videos of setups, changeovers, and cleaning procedures.

Keep the records of machine performance up to date and in a central location. There are good computer programs available to track machine performance and parts usage. If you make the investment, make sure someone stays on top of it. Ask for one or two machine reports each week to make sure the program is being managed correctly.

Consider assigning your lubrication program responsibilities to one or two people. Choose the person based on his or her "language" skills with the equipment. If you have a large facility, divide it into smaller lubrication areas so the people stay with their equipment. The lubrication specialists will visit many of the critical areas of the process each day. They will very often hear the "foul language" before someone else and correct it before it causes a bigger problem.

Rotating maintenance people to different areas of the process has its advantages and disadvantages. The biggest advantage is the development of good maintenance people who truly become the generalists you desire when utilizing this approach.

However, the weaker mechanics can get farther behind and will rely on the better people to carry them. Switching people often, such as every day or every week, does not allow the maintenance specialist enough time to learn the language in a particular area of the process. If you move people around once a month or less, you have a good chance of identifying the maintenance person's weakness and correcting it with a targeted training program. Better training translates into better maintenance specialists, which translates into better preventive maintenance.

Training the machine operators to perform all of the maintenance on the equipment is generally not a good idea. The level of knowledge required to be a good mechanic or electrician is much greater than the level required to be a good machine operator who can also perform some basic maintenance. In addition, the wage rate for a maintenance person is higher than that for a machine operator, because the maintenance person is expected to have skills for all of the machines.

ENVIRONMENT

Finally, who should manage the department? Should the maintenance group report to the process supervisor or to a maintenance supervisor? I have seen both setups, and I believe the latter is the most effective reporting relationship. In a large facility, I saw a very good shift maintenance supervisor managing a group of four mechanics and two electricians become a very average and frustrated shift process/maintenance supervisor. In addition to managing the process, he had to supervise about 30 line employees. The frustration was a result of the realization that neither job was being given the attention it deserved. Resolving people issues will always consume a lot of time.

In small facilities with processes that do not require a large number of machines, the maintenance supervisor can also be the process supervisor. The reverse is effective in very few instances because most process supervisors do not have the skills to be maintenance supervisors.

For medium-sized and large facilities, the process supervisor has too much to manage (the process and the line operators) to effectively manage the maintenance unit on his or her shift. The cost savings gained in the short run from a reduced headcount by combining the process and maintenance supervisory positions will diminish as the process equipment deteriorates from the lack of good preventive maintenance.

MAINTENANCE PARTS

"We need the part today. Why didn't we have it here?" If you have worked in process environments for any period of time, you have undoubtedly heard some form of this conversation at least once. If you are a small, privately owned operation, you probably are already aware of how much you have (or actually don't have) invested in the parts inventory, because one of the most important factors to your survival, cash flow, has taken care of the problem for you. Yet, big or small operations, cash-rich or cash-poor operations have the same question to answer: Do we have the right parts?

The first thing you need is a clean and organized parts area. An organized parts area and maintenance shop improves

PART ONE: KNOWING THE PROCESS

the morale of the people in the department. It also makes it easier to find a part when you need it. So, it improves productivity. It will also impress your boss.

When you get your car fixed, do you get a warm, fuzzy feeling about the quality of the mechanics and their work when you observe a dirty and unorganized work area? I don't think so. You should feel the same way about the maintenance area for your process.

Keeping the shop clean is the responsibility of all of the maintenance people. There should be a schedule that defines who will clean the area and how often it should be cleaned. If you have a multiple-shift operation, make sure each shift gets involved.

With a clean and organized shop, the people can take and keep accurate inventories. The parts inventory should be assigned a dollar value. You can calculate your turns on this inventory by dividing your annual expenditure on parts by the average value of the monthly inventory. This calculation will give you some idea if you have too much inventory. You should look for no less than a 6:1 ratio. This ratio can be misleading when considered alone. If the ratio is 10:1 and you are running out of parts and incurring downtime, you are probably not spending enough on inventory.

Ask your people for a frequency distribution on parts used based on your inventory turns. If your inventory turns are 6:1, ask for the frequency distribution covering a 2-month period.

Compare this usage distribution to your total inventory data. Look for items that don't move. If they are not being used, why do you have them? Can you sell them back to the distributor? Can you use the space more effectively?

Look for the "big ticket" items and note how fast they move. How much of your inventory is tied up in these costly items? Are they essential to your operation? Just because they are expensive does not mean you should carry them in your inventory. Is the vendor close by? How many do they keep in stock? How fast can they get the part for you?

Look at the items that move quickly, as they can provide you with excellent insights about your process.

ENVIRONMENT

What machines or parts of your process are consuming the items? Look at the preventive maintenance program for the machines with continuous consumption patterns. Determine whether the machine needs to be replaced or if the preventive maintenance effort is inadequate. The cost to you in downtime, parts, and maintenance labor may justify the expenditure for the new machine.

The criteria for the purchase and inventory of parts can be developed by separating the parts into four categories: (1) Can it shut down the line, and is it inexpensive? (2) Can it shut down the line, and is it expensive? (3) Is it hard to get (long lead time) and inexpensive? (4) Is it hard to get and expensive? Ask for three to five parts in each category. Compare this list to your frequency distribution and inventory sheet. Look very carefully at any item that is hard to get and can shut down the line. Make sure you have these items in stock.

Once you have an understanding of the parts inventory and how it is being consumed, you can start to work on reducing the dollars you have invested in it. Set an on-hand quantity that is reasonable. Base your reorder point on how fast you can get the part.

Your parts clerk should shop around for the best price. Finding a vendor who can deliver a part quickly at a consistently low price (maybe not the lowest) may be worth more to you in reduced inventory and carrying costs and improved productivity for your clerk. A good vendor can be the same as a "just in time" system for the equipment parts of your process. The vendor can smooth out the effect forecasting errors in the preventive maintenance program for downtime on the equipment.

Here are a few other points to consider. Can any of the parts be made by your people? If you have a machine shop nearby, can they make it? Parts that are expensive and have long lead times can probably be made by a local machine shop. Make sure you use a new part as a model to make the copies. It seems obvious, but I have seen new parts made from worn-out parts and then put into the operation. Because the part may not fit exactly into the machine, the machine is readjusted to accommodate it. A different part for the machine means a different

PART ONE: KNOWING THE PROCESS

effect on the process. You cannot control your process when this happens.

Look at the location of your maintenance and parts departments in relation to the areas of the process where the maintenance people are spending their time. A centrally located shop with one parts area is ideal. Split parts areas can make inventory control difficult.

If these areas cannot be centrally located, then consider using bicycles to move people and parts more quickly. The cost of downtime and maintenance wages will quickly pay for parts carts.

Many computer programs are available that combine inventory control and preventive maintenance. These systems can be an excellent investment if you have the right program to track parts usage and the right person to set up and grow with the preventive maintenance system. You can purchase inventory control software as a stand-alone unit, but tying the information into your preventive maintenance program can give your operation the real home run.

If you cannot afford a computerized system, then a manual card system should be developed to perform the same analyses. The manual system will work, but it will take more time and effort to manage.

ENVIRONMENTAL CONDITIONS

A large number of processes are affected by the conditions in the process environment. Too much humidity can cause sticking; too little humidity can generate static electricity. Variations in the temperature can also affect processes. Each of these conditions affect the people who affect the process by causing them to focus more on the conditions surrounding them than on their jobs.

Understanding how changes in the environment affect your process can be one of the easiest cost savings opportunities you will find. Electronic probes can be connected to computers to monitor temperatures, humidity, wind, and barometric pressure. These devices can continuously record conditions in

ENVIRONMENT

your process environment. You can print a hard copy for your records and analyze the data relative to the performance of your process that day.

If you cannot afford the computerized approach, thermometers and gages with charts are available. This equipment will also do the job, but it requires someone to change the charts, check the ink, and keep it clean and properly adjusted. Information about these conditions is also available at your local airport or in your newspaper, but the actual conditions will probably be different in the process area.

Once you determine what variables are critical to the success of your process, the "fun" starts. If you have the money to buy the equipment to control these variables, it will be a little easier. Make sure you treat the area to be controlled as one room. For example, if similar parts of the process are performed on different levels in the same room, the machines and process may prefer the cooler area. Remember, hot air rises.

Calculate how many times the air should be turned over in the processing room to achieve the desired result. Make sure you do not expose your process to undesired air-flow patterns. Carefully select the appropriate location for your ductwork, vents, and returns. A good heating, ventilation, and air conditioning (HVAC) person can help you here, but do not assume that he or she will have the best solution. You should know your process better and how air movement affects it.

Just because your process environment is already controlled by an HVAC system doesn't mean it cannot be a factor in your process. There are variations in the performance of any system. Is it properly maintained? Check the preventive maintenance sheets for the pieces of equipment that make up the system. Determine whether one piece of equipment has had more than the usual problems lately if you are starting to see changes in the process.

Also, you probably are not the first manager of the facility. The current HVAC may not be the original system. You may now have a different process or product in the room. It would be important to find someone who had experience with each of the changes. A history lesson from the long-term employees can contain the most valuable information you can get about the

PART ONE: KNOWING THE PROCESS

process. If you are lucky enough to find a person with that sense of history, you may have also found someone with the right analytical skills to help see you through some of your current problems.

Finally, even if you are the first process manager in a new facility with a new HVAC, again, don't assume that it can't be a variable. Even though the new system was designed and installed according to a given set of specifications, the actual process may be affected in a different manner. Determine how the specifications were developed.

Visiting other facilities with processes producing the same product can help you develop the appropriate process specifications for your HVAC system. Remember, however, it will not be exactly the same as your operation.

With a new product or process, you need to be aware of the original process parameter assumptions and the specifications. Make sure you are achieving the desired specifications by charting the environment. Test the assumptions against the performance of the process. They were only assumptions, and no one knows the process better than you. If you are lucky enough to be involved in the design phase, make sure you have the assumptions and specifications written down for future reference.

What if you don't have and cannot afford an HVAC system for the process? Nothing much changes. Charting, observing, and analyzing are essential to achieving success in your operation. The difference between the two environments (automatically and naturally controlled) is in the solutions required to keep the process running well. You may be required to modify formulations, equipment settings, how the equipment is maintained, and relief patterns for the people in natural environments.

Some of the changes required to keep you running may have an adverse effect on the quality of the product. You will have to decide whether the product is still in the acceptable range after the change is made. If the product becomes unacceptable, then you should shut down the process.

Once the environment is in control, you will have to reexamine the assumptions about your process. A controlled process environment is a different process environment. You

ENVIRONMENT

may still have to make adjustments to the environment and/or the materials for seasonal conditions.

CHAPTER 3
MEASUREMENT

ACCOUNTING FOR THE PROCESS

"Kill a report and save a tree." If at times you feel you and/or your accounting staff are suffocating from the paperwork burden, take a step back and look at the whole process. Recognize that there are two types of reports: corporate reports and in-plant reports.

Very often, you have no control over corporate reports, but try to make your point about the value of a corporate report if you feel strongly about it. One boss I had was so sure about the marginal value of a corporate report, he stopped sending it in and waited for someone to notice. It was more than 4 months before someone called. Even though it wasn't sorely missed, it was another year before the report was dropped.

A good place to start is with the reports your process generates. Are they redundant? Do they sit in file cabinets? Are they hard to read and fill out? Are half of the spaces not filled in anymore? If the answer is "yes" to any one of these questions, then it is time to eliminate, consolidate, and/or redesign the forms.

How? First, understand how the numbers move from the process environment to the office. How do the numbers get on the first report, and how do they look on the last report? You cannot change anything, whether it is a process variable or a

PART ONE: KNOWING THE PROCESS

piece of paperwork, without understanding what it is supposed to accomplish.

Ask yourself the following question: Does the report or the data on the report add to your understanding of the process? If you understand the process, then you should know whether the data are adding value to your information base. If you're not sure, then wait until you know more about the process. The paperwork system was working before you got there; it can wait a little longer until you're sure.

Once you are sure, don't make the changes alone. Get everyone involved. Ask for feedback from everyone from the line operators to the front office. They will undoubtedly have suggestions, and they will keep you from eliminating the one number on the one report that someone needs in the front office or at corporate.

Another factor to consider is the time devoted to completing the reports. Usually, a fair amount of time is required to collect data, but very often, less time is spent analyzing the data. This productivity paradox can be resolved in part with the use of computers to collect and generate the reports. The time saved creating the report can be devoted to analyzing the report. But who can really determine what is going on from a report? Only someone who understands the process and the numbers behind it.

It is very important that the process accountant understand what the numbers mean for the process. General accounting provides the skeletal structure for running a business, but the cost accounting is the life-blood! It happens too often that the process accountant does not see the relationships between the report he or she is generating and the real world of the process environment. This person is probably spending more time than anyone else looking at the data from the process. The accountant sees the numbers first and should be able to spot a trend quickly. Identifying a trend quickly can save the company money and keep the process going in the right direction.

Encourage your accountant to spend time in the processing environment. Have him or her run a shift for a day or a week. Spend time explaining how you think the process is

MEASUREMENT

supposed to work as you go over the numbers together. It is likely that both of you will learn something new about the process during this exchange of information.

A sure sign that the accounting function is not involved in or does not understand the process is when that department asks process people to perform accounting tasks. Process people should spend their time controlling the process, not filling out forms for reports. Make sure the lines of responsibility are clear and that each group is supplying the appropriate information to the other group.

There are many ways to collect data in a process environment. You can put terminals in the processing environment to collect case counts, temperatures, humidity, machine downtime, and formulation data. Use electric eyes or bar code and magnetic strip recorders for payroll, receiving, shipping, order tracking, and case counts. Collecting, recording, and analyzing good information more quickly will pay big dividends in your quest to understand the process.

If you had no computers, no reports, and no accounting system, how would you know how well you were controlling costs? The old-fashioned way—you would spend time in the processing environment. It still is the best way to get a handle on how the process is performing. The MPPY rule for control is as follows: *Walk, watch, talk, and listen to your process each day.* Be on the floor and not in your office, ask the operators about their jobs, watch maintenance take a machine apart, and know every aspect of the process intimately.

Use the daily efficiency report to guide your "walk, watch, talk, and listen" activities. Ask everyone why the line ran well yesterday. Show them the report. As you learn more about the process, the daily efficiency report will become meaningful to you and the operators.

Ideally, the report should be on your desk by 9:30 AM each day. If it isn't, find out why not and what it would take to have the report by 9:30 AM. Maybe the operators are not providing the office with good working documents of the previous day's activities. Maybe there are too many documents to consolidate. Maybe a computer program to perform the calculations should be developed.

PART ONE: KNOWING THE PROCESS

The accuracy of the daily report depends on the accuracy of the cost accounting standards in the process. Make sure your process standards are accurate. Having "fudged" factors in your accounting system is not consistent with achieving your goal of understanding the process. The performance numbers have to be clear, concise, and honest for everyone to know how well the process is doing at any point in time.

If you change a standard, make sure that you change it for a good reason. Standards are set to measure performance and determine product costs. Marketing and sales departments use this information to set the price-value relationship in the market. Without accurate cost information, your process could be losing sales.

If you change standards too frequently, you will lose track of your performance base, making it very difficult to maintain good budget data and to hold managers accountable for their performance. Managers and operators need targets to improve their performance. As you learn more about your process, hopefully you will change the standards in a positive direction only.

Another cost-control suggestion is to sign all purchase orders. See who is buying what and how often. It is good to give people authority to make buying decisions, but if something is not clear, don't hesitate to ask a question.

Ask your supervisors to initial the time cards or a labor report from their group each day. A labor report (generated by an electronic time-keeping system) is better because it summarizes a group's activities quickly on one sheet of paper. Signing the time cards each day will reduce payroll errors. A supervisor cannot remember on Monday morning what happened each day of the previous week.

Another suggestion is to zero out an item in the inventory anytime you can. If you run out of something in the middle of the accounting period, zero out the inventory at that time; don't wait until the end of the month. "Hold" the truck with the replenishments until you figure usage to that point. The scrap and usage data on that item will be more accurate. The supervisors, the line operators, and the accounting staff should perform live audits on material usage regularly to keep the standards and the process on the same track.

MEASUREMENT

When it is time to conduct an inventory, have a written plan that the people can execute. Prepare a procedures sheet that explains how to collect and record the right data. Use a drawing of the facility to label the areas and items to be inventoried. Use two people for each count team to reduce the chance of error. One person can count and one person can write. Have one audit team follow behind the counting teams to check the accuracy of their counts. Try to have the same people count at each inventory.

If you maintain good standards on usage, have good control over your accounting system for receipts, and have the process in control, then quarterly or semiannual inventories should provide the level of accuracy you need. If there are specific problems, you may want to track the movement on one or two items weekly or monthly until they are corrected.

There will always be room to improve the process and opportunities to reduce costs. You should be looking for these opportunities, but spend your time looking in the right places and at the right things.

Many process managers look first at the labor in the process to reduce costs. Undoubtedly, there will be some ways to reduce the labor in the process by working smarter, improving the scheduling, and replacing manual tasks with machines. The most effective way to control and reduce costs, however, is to control the process. A process that is out of control will consume more money in materials, scrap, equipment repairs, and downtime than the labor you saved by eliminating one line operator.

Look at all of your costs. Don't assume or manage any cost as if it is a fixed cost. Negotiate a better lease. Get a volume discount from the local utilities. Shop around for a better insurance policy. Check the prices of the different scrap haulers.

Look at the variable costs for material consumption as a constant and daily challenge of trying to make more product for less each day. Reduce scrap whenever and wherever possible. Recycle or sell the small amount of scrap your controlled process is generating.

Determine whether the process can reuse some of the scrap by blending it with new materials. If the process is too

PART ONE: KNOWING THE PROCESS

sensitive and only small amounts of scrap can be recycled, then it might not be worth the cost of rehandling and storing the scrap.

Any point in the process in which scrap is generated is a great place to start analyzing and fixing the process. Conduct a scrap analysis of the entire line. Determine the percent of scrap generated at each point and the cost of the scrap at that point. Scrap generated farther along in the process will be more expensive because additional processing costs have been applied to it. Identify which scrap generators can be fixed quickly and at minimal cost and fix them.

A point in the process that is generating a large amount of scrap is indicative of a problem with a very important process parameter being out of control. It may take more time and money to fix this kind of problem. It may also be caused by something very simple that everyone has overlooked.

Encourage people to pick up washers, bolts, and tools that have been dropped. One stainless steel washer is worth about 5 cents. If you saw three or four nickels on the floor, wouldn't you stop to pick them up? At a process meeting, sprinkle the floor with nickels and washers. Walk into the room after everyone has arrived and see what they picked up. It can be a great way to make a point.

Finally, use the 80/20 rule—the Pareto principle—when you are trying to understand what you should tackle first. As your process improves, the "big fires" (80 percent) will get put out. Concentrate on fine-tuning the items that make up 20 percent of your costs after you really understand the process. Items in the 20 percent category are the subtle aspects of the process that will be hard to identify and resolve, but they are there to discover. All you and the operators have to do is "walk, watch, talk, and listen" to the process every day.

COMPUTERS AND STATISTICS

I tell my friends that proof of our success as a civilization can be found in three inventions: indoor plumbing, automatic teller machines, and hand-held cordless vacuums. Could we do

MEASUREMENT

without any one of the three? I am sure that each of us could come up with our own list, but computers are sure to be on every list.

Most companies that are part of a corporation are tied into the mainframe computer at the home office or some central location. Payroll, distribution, and sales orders are the most common centrally controlled programs. Purchase orders for equipment, parts, and supplies are the most common locally controlled programs.

Very often, you or your people will feel frustrated and alone at the end of the "mainframe computer rope" when problems occur or when the program is not the most user-friendly. But don't feel like you are powerless. If someone has a suggestion about how to improve a program or how some time could be saved if the computer could give some of the data back in a different form, listen to them. Call some of the other process managers and discuss the idea. They may have the same needs, and you can work together to improve the process.

Every day, more and more people are becoming computer literate and therefore may have some good suggestions. (Why doesn't anything else sound as nice with the word literate? Auto-repair literate, plumbing literate, and managing literate?)

If your process doesn't have a computer yet, then base your decision on which one to buy on the software you might like to use. If you are buying more than one or plan to in the future, look for a type of machine and software that can be networked.

Use the personal computer programs to help you improve the productivity of the corporate programs. Very often, corporate software is chosen for corporate goals. These information goals may not be compatible with the way information is assembled at the process level. The person entering the data into the computer may be spending 1 hour each day preparing the data before they are transferred to the corporate computer.

A personal computer with an electronic spreadsheet can summarize the process data before it goes to the mainframe computer. It might be possible to transfer this information electronically between the computers. If the data have to be

PART ONE: KNOWING THE PROCESS

manually keyed into the corporate program, then design the spreadsheet so that the information is listed in the same order. Keeping the formats the same for data entry will minimize keypunch mistakes.

If it is possible, volunteer to be on the steering committee when new software is being purchased. Knowing how your process handles information and what kind of information it needs can be of great value when selecting the right software package.

Visualize your long-term plan and explain it to a couple of vendors. They can be very helpful, because they are knowledgeable about the newest developments in the field. Look at the ability of the company to provide technical assistance and repair service before you make your purchase decision. Make sure they will be around for the next few years. People, software, and hardware change faster in the computer business than diapers on a newborn baby.

Accurate and timely information on the performance of the process is essential to the development of an attitude of continual improvement. Good process decisions require good information.

What is good process information? The best kind of process information is based on "the facts." The best kind of information is information collected and analyzed with statistical tools.

In the 1980s, the use of statistics in process environments increased significantly. There were little armies of people scurrying about in many facilities plotting control charts, taking samples, and measuring ranges. This change was very good for operations people and there is no turning back. If you still have not learned about statistical process control, analysis of variance, or quality loss function methods, start immediately.

It is not essential that the process manager collect the data, but he or she should understand how the data are collected, what it means for the process, and how to ask the right questions about the data. Ideally, you can identify someone in your operation with some statistical ability and good communication and analytical skills to be in charge of the program.

MEASUREMENT

The entire work force should be exposed to the concept, but it is not realistic or cost-effective to expect everyone to become experts with the statistical tools. Concentrate your training efforts on a group of people specifically selected for their exploratory and statistical abilities.

Small groups of four to five people can work wonders with a little direction. Start on one shift and move to the other shifts when the people are familiar with the concept. You should have someone with experience from the other shifts in the first group on the new shift. The use of statistical process control charts will provide you and the line operators with great insights into the how's and why's of your operation.

If you haven't taken a course in statistical process control, then use frequency charts, which can reveal trends in the process that you may not be aware exist. Watch a part of the process several times at random each day for 20 minutes. Record all of the reasons scrap was generated or the machine broke down. In a very short time, a pattern will develop with the data, and you will have an opportunity to correct a process problem.

All of the data in the world will be of no use to you or your operation if you don't take the time to analyze them. Make sure time is available to review and discuss the data.

Many books and seminars are available on the different statistical models that can be used in a process environment. If you have a computer, software is available that automatically produces the charts once the data have been entered. You still have to figure out what it means to your process, but the software makes it easy to manipulate the data and ask the "what if" kind of questions about the behavior of the process that lead to improvement breakthroughs.

One big advantage to using the computer is the ability to save the database and to compare it with a new one after a change has been made. It will also allow you to develop a large database over a long period of time to spot more subtle trends. Once, we compared three identical machines on the same line over 6 months and saw a seasonal trend we had not noticed. It consisted of more than 2000 data points for each machine. Without the computer, it would have been impossible to manage that number of data.

PART ONE: KNOWING THE PROCESS

An experienced manager can get an intuitive sense of how well things are going by walking around and observing the process each day. Subtle changes and opportunities, however, may not be visible. The process and how it is interacting with equipment is too complex to be analyzed by the normal human senses. The process equipment has become too complex for a person to rely solely on his or her intuitive senses. You need the statistical tools to monitor the process effectively.

The subtle changes a statistically based program will detect are the areas of improvement that will move the process into the 90 percent efficiency range. These changes can be observed in the routine activities that are performed each day by the line operators and the machines. Some of the discoveries will change those "because that's the way we've always done it" approaches to the process. Let the data and the science of the process they are illustrating do the talking as you move the people and the process along the road of continual improvement!

PART 2
KNOWING THE PEOPLE

How is your organization structured to support the process? Do the different functional areas understand how their specific, specialized skills and activities affect the performance and quality of the product being produced by the process? Do the different functional areas understand how their skills and activities must interplay and overlap with the skills and activities of the other functional areas for the entire process group to be successful? Do you understand how to bring each of the units together in your effort to continually improve the process?

Managing the technical aspects of the process requires the use of analytical and problem-solving skills as you attempt to understand the science of the process. These skills must be combined with the ability to communicate with, teach, and motivate the people working in the process. Improving the process is contingent on creating an environment in which the people can contribute in an enthusiastic and productive manner to improving the technical aspects of the process.

Most process environments are supported by similar departmental functions, independent of the nature of the process. The success of these departments is determined by the people principles and processes that have evolved within each department. Some of the people principles are as follows: How do the operators feel about their contributions at work? How do they feel about the company's future? How do they communicate with each other and with their supervisors? How do they resolve process issues? Do they feel involved in the effort to continually improve the process?

The success of the process is directly related to identifying and understanding the sociological and psychological dynamics of the process group. The soft, human issues are often more difficult to deal with than are the pure science-related issues. The people variables, such as being tired or sick or having family problems, will consume a great deal of your time and energy. The human issues will affect the process in ways that cannot be documented or measured. At times, it will seem as if these issues are all you are working on. Yet if the time and the energy are focused on the right issues, the return on your investment in this area of the process will easily justify itself. Learning how to manage the wide variety of people activities is the second important element in your efforts to "MPPY" the process.

CHAPTER 4
THE SOFT ISSUES

QUALITY AND THE PEOPLE

People want to work in a quality environment that produces a quality product. The product does not have to be the most expensive product of its class to be a quality product. Even the most inexpensive candy can be produced in a quality environment to the highest quality specifications by people actively involved in the process. People want to take pride in their work efforts.

How do you get your people involved? How do you get your people to care? How do you develop pride in your people? The answer to all of these questions originates at the same place in your process. It begins with you. It requires an honest commitment by you to achieve and maintain a quality work environment. You have to be proud of the process, its products, and its people.

If you don't truly feel this way, your employees will realize it. When they sense that you don't care, they will develop an "I don't care" attitude as well. Your heart must be in it. Token gestures toward quality in any form will not have a positive effect on the people and the quality of the product.

Once you make the commitment and have the resolve to stick with it, don't look for immediate returns. This kind of

PART TWO: KNOWING THE PEOPLE

effort has a "snowball" effect and a cumulative payoff. The more people believe in you, the more things will improve. The degree to which your plant has swung to the "dark side" in quality will determine how quickly it can recover.

The quality commitment is one of the most interesting management concepts you will encounter. Why? Because you can't go halfway; it's all or nothing. There is no gray area.

The people can recognize the depth of your commitment; if you waiver or backtrack from that commitment, look out! Once you take a step backward, or hedge in the slightest way, the next step forward will be very difficult. The people will be reluctant to accept it as a renewed commitment. They will return to feeling like an unimportant cog in the process machine. The line operators will question their ability to have a positive effect on the process. They will feel that their contributions will not be appreciated. The degree to which the line operators adopt these feelings will depend on the degree to which they feel they have been duped. Their reaction is a simple defense mechanism to protect themselves from getting hurt, embarrassed, or taken advantage of.

When you have made the honest commitment, the next step is to get the people to do the same. This effort starts with a belief that everybody contributes to the success of the process. Every person in the process helps to make a quality product. Everyone has a role to play in the process scheme. If one person doesn't contribute, doesn't pull his or her weight, it can and will affect the entire process. The team is a unit, and the unit is a team.

Here is a story that many of you have heard in one form or another, but still it illustrates the team concept in effectively creating a quality process environment. A new employee was hired into a facility and assigned to the sanitation crew. His first assignment was to clean the washrooms and locker rooms. He was performing the task, but the assignment did not fit with the grander illusions he had entertained when he was hired at the facility.

As it happened, he had some help in acquiring the position—his parents were the friends and neighbors of the process superintendent. Joe, the superintendent, walked into

THE SOFT ISSUES

the rest room and saw the young man. He asked, "How is it going?"

The young man replied, "Well, not too bad, but I didn't expect to be cleaning toilets."

Joe said, "Tom, would you like to be the superintendent?"

"Absolutely, but I realize I have plenty of things to learn first," said Tom.

"Yes, you do," said Joe, "and the first thing you need to learn is how to clean a rest room better than anybody else ever has at this plant. Then someone might notice your efforts. But it won't be easy."

"Why not?" Tom asked.

"Because I still think I cleaned this washroom better than anybody has since I've been here. It was my first job, too," said Joe.

A quick inspection of the rest room area will tell you what you can expect to see in the remainder of the facility. This simple task is just as important as the other tasks in the process. It will provide you with a good idea of the level of quality commitment in the plant.

People armed with a belief that you care about quality and a confidence in their ability to contribute to improving the product quality can be a great source of information about the process and its quality parameters. They will start asking themselves, each other, and you the following questions: Why did this happen? How does that work? Can we make it better? They will get into the "nuts and bolts" of the process.

They will want and expect you to get into the process. They will expect you to start asking them *why, how,* and *can?* It is amazing how many managers don't bother to ask the line operators for their opinions on an issue. The combined experiences of the operators can be invaluable in figuring out a process problem. You may be just what is needed to bring the analysis together.

You will have to respond to their challenge to be involved, because sooner or later the line operators will ask the question that makes uncommitted, uninvolved process managers cringe from their lack of knowledge about the process. The same question makes committed process managers tingle

PART TWO: KNOWING THE PEOPLE

from the excitement and possible success and fear of how to accomplish it. The question is *when?* When can we do it? When can we change it? When can we make the product and process better?

You have to be involved in and knowledgeable about the process to be sure the line operators' suggestions are correct. You will have to judge whether the investment is worth the time and money. Is it a real opportunity or just a small step in the right direction?

Is the suggested improvement a red herring or a blind alley? You may decide to let the line operators test the idea for the learning experience, even if you know it is a "red herring," to keep their enthusiasm for continuous improvement growing. Chasing red herrings can be costly and may affect the current level of quality, so be careful. Understand what the line operators can learn and the implications for the process before you decide to let them experiment.

It may be difficult to implement a costly idea immediately. Most process managers do not have a blank checkbook. If the project looks expensive, don't stop there. Remember your commitment?

Identify all costs. Determine whether the project can be performed by the people in the facility. Ask for help from vendors, if necessary. Identify all of the opportunities. Present the complete analysis to your boss. Remind him or her that when you embarked on your mission to improve morale and quality and to make this the best process in the system, you knew some extra funds might be necessary. Demonstrate how the investment will pay for itself in improvements to the process.

To get something, you have to give something. First, you can give your boss some improvements in the process that didn't cost anything. There are plenty of low-cost or no-cost improvements in every process waiting to be discovered. Your boss will feel better about investing in more costly projects to improve the process if he or she is confident that you understand the science of the process.

Discovery is the phrase or feeling that really expresses the essence of the correct attitude toward continually improving the process. As you and your people delve into the process,

THE SOFT ISSUES

you will feel like explorers on a discovery mission. You will feel the same thrill when that special moment of insight occurs. Recognize the fact that some explorers are better than others. Find out who the best explorers are in your facility and keep them involved. Test all of the discoveries in all parts of the process against their exploratory instincts. A different viewpoint can provide good insights.

You will have to deal with some "nonexplorers" and their impact on the process and quality. Turnover in some of these employees may even be required to achieve the ultimate goal. Changing employees can take time. Even though there are usually only a few "bad apples in the barrel," targeting specific training improvement programs for these individuals can have a dramatic effect. These people can be mavericks or leaders, but try to help them improve with training or counseling or help them leave. Most of the time, the people will improve and thus the quality of their work and that of the people around them will improve.

Some type of laboratory is beneficial in a process environment. It can provide a home for some special test equipment, if the process environment is too harsh or if the tests require chemicals or other items that cannot be placed in the process environment.

However, if the test equipment can be placed in your process environment, then it should be. Let the people perform the tests and discuss the results and possibilities. This step helps develop the explorer attitude in more people, because more people will be discussing process issues on the floor as they are occurring.

Make sure your supervisors work through the tests and the results with the line operators. Request summaries of the tests for review. Every now and then, conduct a test yourself. Develop homemade test equipment if you cannot purchase the proper equipment.

Every aspect of employee involvement in the quality program takes time. Make sure the line operators have the time to perform the tests properly. I was familiar with one situation in which the quality department was dissolved, and the tests were moved to the line function. The operators were not

PART TWO: KNOWING THE PEOPLE

able to perform the tests correctly, resulting in a deterioration in the product quality. The attitude of the line operators deteriorated with the quality, because they felt it was just one more thing for them to do in their already busy day. Don't go halfway on the quality commitment. Tests take time. Make sure time is available, and make sure the test is worthwhile.

Finally, the people must be thoroughly trained in the test methods. Explain everything about the test. Why is it being done? What is it supposed to measure? How are the results supposed to be interpreted and applied? Where and how often are the results supposed to be recorded? Invest in the training of the operators and the supervisors to avoid mistakes later in the program.

After 2 or 3 months of working with the program, get the group together for a meeting about the procedures and tests. They will probably have some very good insights into the testing methods and the process and how to improve them.

If you walk into a new assignment with a well-entrenched lab, take your time (if you can) in changing it. Many labs become great caverns for expensive equipment, costly tests, and file cabinets full of unanalyzed process data. Ask to see all of the reports and paperwork the lab personnel are currently working on and filling out. Become very familiar with all of the tests. Use this knowledge to solidify your knowledge of the process. Once you are comfortable with your knowledge of the process, eliminate and consolidate those reports and tests that do not add value to the quality performance of the line or the knowledge of the process.

A good place to start is with those reports that do not make it back to the line supervisors and line operators. Why? Because the operators are the people who most need feedback on the process; if it is not getting to them, you need to find out why. Maybe the feedback can help them perform their job better. If they cannot use the feedback, then you have to determine what value it is adding to the process and whether the information should still be collected.

The reports going to the line operators can also be a great tool for you to use to find out who really understands the process. Ask around about the data on the reports to identify the explorers in the process. Find out which people are using

THE SOFT ISSUES

the data and how. Find out if everyone is using the data the same way. Maybe the reason for providing the data to the line is no longer relevant, because the general knowledge level of the process has gone beyond it. Maybe the data still have value, but nobody ever explained it to the line operators.

Include the line operators in the meetings when you review your product quality and compare it with that of your competitors' products. You should look at your competitors' products frequently. Look for any kind of change. It might be a signal that they are onto some new great idea or that they are catching up. If you truly understand your process, you should be able to detect changes in competitors' product and relate the possibilities to your process.

COMMUNICATING WITH THE PROCESS EMPLOYEES

For all of the communication on communication, managers in process and corporate environments usually do not get high marks from their employees for their communication skills. There are many different programs, but the most successful communication programs have four characteristics in common. Success was not a function of the number of group or individual meetings. It was not the number of memos or colorful charts on the bulletin boards. It was not the number of productivity contests or suggestion boxes in the facility. Good communication was a function of the honesty and accessibility of the management. The MPPY rule for successful communication is as follows: *HAVE it! You must be Honest, Accessible, Visible, and Enthusiastic.*

Honesty means treating people with respect. One of the best pieces of advice I ever received was given to me by the vice president of manufacturing as I was about to leave the corporate office to embark on my first assignment into a production process environment. He said, "Remember that these people are no different than you when they leave the facility. They have to feed their children. They have to pay their bills. They have happy days and sad days. They have successes, and they have problems."

PART TWO: KNOWING THE PEOPLE

It sounds so simple, yet it is so quickly forgotten by many managers when they sit down behind their desks. Being honest means being honest with yourself. You have to accept the people who work for you as equals and recognize them as individuals.

When you talk to people, use their first names. The use of first names or nicknames makes people more comfortable and relaxed. Name tags are helpful.

I was in a facility of about 300 people, and the only people with name tags were in management. I think it was a cost-saving idea. Besides being almost impossible to remember everyone's name, think of how anonymous it made the line operators feel.

Communication with the process employees on a first-name basis breaks down the defensive, impersonal barriers people erect in work situations. Another MPPY communication rule is as follows: *Recognize that work is a social situation.* People get together for a particular reason and interact to try and make it go smoothly.

There are certain socially acceptable forms of behavior people adopt at a party and at work. People are polite. They give each other directions. They find someone to talk to. Working in a process environment is like going to the same party with the same people 5 days each week. Learn the first names and a little bit about the families of the process employees to help you become the life of the "process party"!

Accessibility means letting the people see the real you. Letting the people get to know you is an important ingredient in becoming a good process manager. Let the real you be accessible. It is acceptable to show anger, joy, and sadness. Maintaining a stiff professional image all of the time is not going to earn the respect of the employees. You are human just as they are. They know it and expect to see that side of you. Some professional distance is appropriate, but it should be viewed from the people's point of view as an element of your leadership qualities, not your invulnerability.

People do not have to know everything that is going on in the facility; they do not expect to know everything. However, they do expect the management group to be aware of everything that is essential to the survival of the process.

THE SOFT ISSUES

If the management group is not aware of something, the line operators can accept it, but they will want some appropriate response once you become aware of it. The people will judge you and your effectiveness by your responses to every situation. Being accessible will help the people understand your responses, because they will understand you.

The belief that the level of knowledge increases the higher you rise on the management ladder is found in every organization. The interesting paradox is that the knowledge of the science of the process an upper level of the organization utilizes to guide the business is quite limited. In reality, the specific knowledge that generates the product and determines the success of the company is found at the lower levels of the organizational chart.

The survival of the process and the company is contingent on the direct and indirect relationships these levels establish with one another. The success of the direct systems, which allows the people to function efficiently each day, is a result of your accessibility. The faith the people have that the indirect system of "invisible accessibility" is working is also a result of your accessibility. The invisible system implies that everyone from the president down is willing to listen.

Visibility means being out of your office and on the floor. You cannot talk to people from your office. Walk through all areas of the process each day. You don't always have to stop at every point in the process. Maybe you just need to stretch your legs or to think through a paperwork problem. Seeing you several times a day reinforces the perception of honesty and accessibility.

A good indicator of how honest and accessible you are is if people come up to you just to talk. It shows that you have effectively broken down the barriers of authority. Always respond in a positive manner to this type of communication.

You should also be prepared to hear about the employees' personal problems with their spouses, children, and families. I have had discussions about divorces, affairs, sex, and college education with employees. Be careful when this happens. For the most part, people just want someone to listen. Because you have been successful in being honest and accessible, you have earned their trust as an impartial friend.

PART TWO: KNOWING THE PEOPLE

The employees may also ask about an improvement to the process they were expecting. This kind of question is also a good sign that you are honest and accessible. They believe you can do something about it. They haven't given up on you or on improving the process.

Always respond to what they are saying. Get back to them later with more information if you don't have the best answer at that time.

Enthusiastic means being excited about the products being produced and the manner in which they are being produced. The people should feel proud of their contributions to the process and the profitability of the products. You should remind them of their important role in the success of the business unit. You have to become the No. 1 cheerleader of every person's efforts. When you are honest, accessible, and visible, it is easier to be enthusiastic because you know how the people feel about the process. You know that they know that you know what is going on.

Being enthusiastic about the process also means being supportive when a suggestion for improvement is made. It may not be the best idea and it may not even work, but the next idea the person thinks of may be excellent. Respond in a manner that allows the person to be comfortable enough to suggest an idea that may not work.

When you communicate with the process people, be enthusiastic in your tone, even if there is a negative event or a situation you are trying to correct. Emphasize what has been learned and discuss the appropriate steps that will be taken to prevent its recurrence. Most employees do not intentionally make mistakes, and there is nothing positive to be gained in the long term by belittling them.

Some situations will require a "heavier" discussion with an employee or supervisor. This type of discussion should never take place in front of other people in the process environment. Try to end any private discussion on a positive note by pointing out the value of his or her contribution to the process.

Avoid extended discussions when a line operator is critical of another manager. Indicate that they should follow up on the issue with the manager.

THE SOFT ISSUES

Show support for the staff but determine whether the questions a line operator raised about a manager are valid. You may have to provide some additional training or have a talk with the manager. The problem may also be the result of a personality conflict. Be careful and determine exactly what is going on before you make a hasty decision. Be guided by the big picture. Everyone, including managers, has bad days and makes mistakes. When was the last time you liked everyone you met at a party?

How do you know that what you are saying and to whom you are saying it is being heard the way you want it to be heard? It brings to mind the old game of having a story told to one person and having him or her repeat it to another, and then another, and so on, so that after it goes around the room and is repeated aloud, the final version is usually quite different from the original story. The best insurance you have to make sure your thought was communicated and heard correctly is to repeat the thought with the same or slightly different wording. Repeat it, but emphasize different parts of the thought with different tones. Good teachers and speakers frequently use this technique.

Repeating a point and reviewing it at a later date increases the chance that the person will commit it to his or her long-term memory. Always ask the person or group if they understood what you said. Repeating a point also helps you make sure that you said the right thing in the right way. Very often in this type of informal discussion, you will find the key phrase that captures the essence of what you are trying to express. That is the phrase to repeat.

Use pictures, graphs, videos, or anything else that can help you enhance the efficiency of communicating the idea. But remember to be concise and to the point. Designate bulletin boards for certain types of communication so people can focus where you want them to and filter out the unnecessary information. If necessary, have a safety board, a schedule board, a quality board, and a personnel board.

Communicate to the managers on a professional level but also be personable and accessible to them. I once attended a staff meeting in which the manager of the process replayed a

PART TWO: KNOWING THE PEOPLE

cassette recording of his observations during the morning tour for the line supervisors. He would turn it off after each observation and wait for the response by the appropriate manager. Why did he have to attend the meeting? He could have just let the machine run. It would have been just as impersonal but much less intimidating. If you use a recorder, take 5 minutes before the meeting to transcribe your discussion points.

Consider a simple thing like turning your desk around to make the line operators more comfortable when you are talking to them. A desk is a sign of authority, and sitting behind it increases the perceived distance between you and "them." Why hide behind it? Besides, it will make your office seem bigger.

Many organizations use recognition programs to identify, reward, and communicate the positive events in the process environment. These programs range in cost and depth from a simple photograph and a name on a plaque to cash rewards based on a percentage of the projected savings. Some form of recognition is always positive. A personal thank you and some conversation can be appropriate. Most of the time, people don't expect anything more. If you have created a friendly, stable, fun working environment, then probably nothing else is required.

Programs that reward people with money are difficult to justify. Most companies do not have the resources or the desire to share those resources with employees over the long term. Environments utilizing this approach are usually looking for quick fixes. After a period of time and improvement, the program is scaled back or stopped. What kind of message does this send to the employees? Don't be creative, problem-solving explorers unless you get rewarded!

The goal of any reward program is to generate ideas that improve the process and communicate the successes. If you have earned the respect of the employees, and they believe in your commitment to the process and their livelihoods, you won't need a special reward program to improve the process. The people will tell you what they think—just ask them! In the best situations, they will tell you without your asking and will already have someone fixing it.

THE SOFT ISSUES

Achieving some milestone can be cause for celebration. Maybe there was a new efficiency record, a new safety record, or a significant improvement in quality. A simple reward, such as free coffee or pop for a day, can show that you appreciate their efforts.

The important thing to remember if you do decide to reward people is to keep it special. You know a reward program has gone stale when you hear a comment such as "What is this for?" Talk to people on the different shifts and see if they know why the reward is being given. If they do not know what is going on, then you are not communicating your successes effectively.

Confidence and enthusiasm are built on communicating the successes the group is achieving. Don't miss the opportunity to say thank you for a good job.

Finally, the process and its different components will be managed by written, scientific, statistically oriented polices and procedures. Other parts of the process will be managed by informal procedures. An MPPY people rule to remember is as follows: *All parts of the process are managed by the people.* The success of any process is directly related to the communication patterns you establish to help the people manage the process.

UNION VS. NONUNION

The ability of management and the employees to communicate effectively can be measured to some extent by examining the role of the union or nonunion relationship that exists in the facility. A positive, productive dialogue can exist between the groups in a union or nonunion environment if they feel that the communication patterns are honest, open, and mutually beneficial. Problems usually surface over the definition and the short- and long-term effects of mutually beneficial issues.

The Wagner Act was passed in 1936 to protect the rights of the employees. This piece of legislation forced employers to consider the long-term effects of their decisions and to acknowledge the right of the operators to participate in the communication patterns of the business. The product of this

PART TWO: KNOWING THE PEOPLE

legislation is paradoxical in light of the efforts of many companies to actively involve the operators in the decision-making process.

There is still a need for some form of legislated protection for the employee and the company. The National Labor Relations Board and the Occupational Safety and Health Administration (OSHA) have mechanisms for these purposes. They may not be perfect, but they do provide a framework within which each group may work.

The type of environment that is the best for the process is very difficult to determine. Many companies have worked very hard to keep their processes nonunion. Many companies have also closed down and/or relocated processes to nonunion geographic areas rather than enter into a business relationship with a union.

Other companies in the same geographic area with basically the same types of employees have efficient processes and produce quality products in a union atmosphere. The key to success in both environments is communication and a strong mutual desire to succeed in the long term. In either situation, you must understand the position of your company and work within the framework of its policies. It is too complex a subject for you to deal with alone.

Clearly, the trend in the last 30 years has been away from union environments. People feel more comfortable with the benevolence of today's employer and the level of legislated protection. Union membership is also declining due to the changes in our work force, from manufacturing to service-oriented jobs.

A problem in a nonunion plant is treating the employees too "carefully." This philosophy can affect what should be a good business decision when evaluating the performance of an operator. People in this situation may be less likely to be held accountable for their actions for fear of upsetting them. The threat of a union vote is omnipresent, and that fear, incorrectly, becomes the basis for managing the activities of the process. In this type of environment, a union may be better, because it defines issues more clearly for both sides.

In nonunion settings, there is usually a document outlining company policies and procedures. This document is similar

THE SOFT ISSUES

to a union contract. Why? There still has to be something written that allows you to run the business in a fair and professional manner. Whatever kind of document exists in the facility, get it, read it, and digest it thoroughly.

If the process is nonunion and you want it to stay that way, attend a seminar on the topic. There are things you can and cannot do if union activity begins in your facility.

If the process is unionized, the same seminar would also be helpful to you. A seminar on how to deal with a strike and negotiations would also be very helpful. Many professionals are available on this subject, and one of them has had experience with a situation similar to yours.

The communication mechanism between the union and management is negotiation. The heart of the negotiation process is the win-win concept. At the end of negotiations, the management and the union both want to appear as winners to their respective constituents. Obviously, in the win-win scenario, each side perceives the result as win-lose. Victory cannot be had without a defeat!

Win-lose, lose-win, and lose-lose situations can create uneasy tensions that will turn the focus away from improving the process. In that situation, no one wins. Without a highly tuned process and the constant devotion to improving it, any result, even a win-win, is meaningless.

Achieving a win-win can be as simple as talking off the record with the negotiator. Discuss the sensitive points between the groups. Suggest ways to smooth them out in your private caucus, and then take the resolutions back to your parties for their review.

President Carter used this method at the Camp David peace conference. President Sadat and Prime Minister Begin could not resolve issues when they were dealing directly with each other. Carter decided to separate them. He worked out their differences during many private discussions with each of them.

When you are negotiating, have the exact cost of your proposals for the negotiations committee to review. Determine the costs of their proposals and communicate the impact on the business to their negotiating team. Always keep the process

PART TWO: KNOWING THE PEOPLE

moving forward. Negotiation requires give and take from both sides.

It is also extremely important to keep the employees informed about what is going on in the facility and in the company when official negotiations are not taking place. Let them know when new equipment is coming and how much money is being invested in the process, their jobs, and their futures. Use bulletin boards, handouts in payroll envelopes, and department meetings to communicate with the line operators.

All of these communication tools can help maintain the proper atmosphere in a union or non-union process. If your intentions are not sincere the people will recognize it.

Simply put, treat people and their ideas with respect. What's the best way to show your respect and communicate at the same time? HAVE it! Spend time in the plant and not behind your desk. Call people by their first names. Ask them if everything is okay. Be prepared to respond if they point out an opportunity to improve the process. The main point is to get and keep the communication ball in your control. Do not let the union, the "grapevine," or the rumor mill be the principal form of communication to your people.

HIRING AND PROMOTING

A considerable amount of time and money is spent on finding the right person for the job. Companies spend large sums to attract the top management required to run their companies. They want to increase the chance of their success and, therefore, the company's success. The search process can take months. The company is looking for the right attitude, the appropriate experiences, and the best fit with the existing culture.

Your boss should have gone through a similar process in hiring you for your present position. Managing a process requires a considerable amount of skill in many different areas. The person in charge is responsible for a significant investment in the building, the equipment, and the company's future profits.

So, you have to ask yourself the following question: How carefully do you select the people to manage and run the process

THE SOFT ISSUES

for you? Your process of selection should be just as thorough. Obviously, you will not have the luxury of waiting 2 months to fill a line position, but you could have a waiting list of pre-screened and interviewed candidates. If you live in an area with a large labor force, this prehiring planning is entirely feasible.

Your ability to hold candidates on the waiting list will depend mostly on the work environment you have created in the facility. Yes, environment, not money, will attract and keep the best people. Many surveys have confirmed that the working environment is more important than money. The wage scale should be competitive with that of other employers in the area that require the same skills, but if your process meets that criterion, getting the best people in your facility will be a function of the working atmosphere. The people in the community know the best place to work, and word will get around.

Hiring should be viewed as a process of selection, not elimination. A specific program should be in place to provide consistency to the hiring process. The program will improve your chances of finding the right person for your process.

The hiring process should start with a written job description of the duties and responsibilities of the position to be filled. It can be brief or detailed, but it should be written down for the benefit of the future employee, his or her supervisor, and you. The expectations for each level of performance are better understood if you take the time to spell them out. The job may have changed since the last time it was filled. The direct supervisor of that area should review the job description to make sure it is still relevant. Identify in the job description the specific process responsibilities associated with the position.

You can improve your selection process by having a brief written test on the process responsibilities. For example, if the position requires an understanding of decimals, fractions, percentages, or weighted averages, a couple of problems involving these elements should be included on the test.

If the operator can use a calculator in the process, then let the applicant use a calculator for the test.

If the job requires reading meters or gages, draw pictures of the instruments or have working models. Ask the applicant to read the settings.

PART TWO: KNOWING THE PEOPLE

If the job requires the ability to understand simple chemical or physical principles, include some questions on these topics.

Maybe there should be a couple of word problems on process troubleshooting. This type of test can be a prescreening tool as well as a pretraining tool. If the employee you select is deficient in only a couple of areas, then you can concentrate your training there.

Once you get to the interview stage, your list should be much smaller. You don't have time to interview 20 people for one position. Limit the interviews to no more than five.

Prepare a set of written questions to ask. The questions should be people oriented and should give you an idea of how well the process and the person will work together. Here are some questions you might ask:

- What are your favorite leisure activities and hobbies?

- What books have you read recently?

- What are your personal and career goals?

- What do you feel are your strengths and weaknesses?

- What did your last boss feel were your strengths and weaknesses?

- What did you like/dislike about your last job?

- What kind of environment do you like to work in?

- How do you react to mistakes by subordinates, peers, or your boss?

- How would you react to a process problem?

- What kind of boss do you like to work for?

- Why would you be good at this job?

THE SOFT ISSUES

Take notes of the applicants' responses. You will want to review them later with the other people involved in the interview process, and you won't be able to remember everything.

Involve other people in the interview. Different perspectives will provide valuable insights.

The most important role you will have to play is that of listener. Your job is to get the prospective employee to talk. You need to learn everything you can in a short time, and you cannot accomplish this goal if you do all of the talking. At a minimum, the candidate should talk about two-thirds of the time.

I once had a 3-hour interview during which I talked for about only 30 minutes! I would have talked even less, but I interrupted the interviewer to ask some questions and to allow him to catch his breath. At the end of the interview, I knew all about him, but he had no idea what I was really like!

If you are momentarily lost for a question, remain silent. Silence can be a very powerful and useful tool to get the other person to talk.

Listen carefully to the candidate's responses and redirect the conversation only when necessary. When you get good at interviewing, the person being interviewed will usually answer several of your questions without you having to ask them.

Listen for certain phrases that would allow you to test the honesty of the answers. For example, if the candidate says "I have known three people," "I have been in two facilities," or "I have worked in a similar environment," ask for the names and locations of the facilities and how they are similar. Sometimes people tend to paint a "pretty picture," and you need to let them know that you are going to test their picture of reality.

It is important to find the person with the right skills for the position, but it is just as important to find the person with the right "chemistry" for your process environment. A person's attitudes, moral outlook on life, and personal problems will have a major impact on your process and on the existing employees.

Is his or her glass half full or half empty? Is the person an explorer? Hiring the right person for the culture of your process is

PART TWO: KNOWING THE PEOPLE

another one of those tasks that, if you do it right, no one notices, because the new person fits right into the process family. But if you hire the wrong person, your abilities will be questioned.

At some point in the hiring or promoting process, you will look at the applicant's educational background. A question will arise: "To be degreed or not to be degreed?" Degrees can be a good indicator of ability and for certain positions they are an appropriate filter.

If your company requires degrees for certain positions and it is inflexible on the issue, then, obviously, you must stay within those parameters. If degrees are not a requirement, then I suggest you be open minded about nondegreed applicants. Some of the best maintenance engineers I have worked with came out of the Air Force. In either case, always, always, always take the time to check out the credentials. Don't assume the candidate graduated from anywhere just because the resume says so.

Every effort should be made to promote from within the facility as this is very good for morale. You and your management group should be developing talent within your facility every day, yet not everyone is promotable and not everyone wants to be promoted. People have different needs and levels of aspiration. The process benefits from the diversity of jobs in a process and from the diversity of people in the process.

Prepare written organization charts for each department of the facility. Identify who can be promoted or moved into key line operator and technical and management positions. Determine who can be moved immediately and who might need some training. Good succession planning will ensure that the process and the facility remain stable over time.

Sometimes, a need will arise that cannot be met from within the organization. Bringing an experienced person into the organization can have immediate benefits, because little or no training is required for the person to acquire the skills needed to fill the position.

It will take some time for the new person to be assimilated into the culture. A new person will also have insights into the culture of the process. After 2 or 3 months, it would be worthwhile to discuss the differences between his or her old and new job

THE SOFT ISSUES

environments and to identify how the differences are helping or hindering your process.

TERMINATING

Terminating an employee due to poor performance can be a long, difficult, and expensive process. Even in situations in which a gross violation of a company policy has occurred (use of alcohol or other drugs, fighting, or theft), termination may not be immediate, as the grievance and hearing procedures could keep the issue unresolved for some time.

The *first* MPPY rule for terminating an employee is as follows: *Spend the time and money to hire and train the right employee.* From the moment you are placed in charge of an organization, make sure you are involved in the hiring and promotion processes. Make sure the new employee passes his or her training period. You may still make mistakes, but your chances of success will be greatly improved.

When it becomes necessary to remove an employee from a position or from the facility due to poor performance, then follow the *second* MPPY rule for terminating an employee: *Go slow and be sure.* At the end of the process, the reasons for the discharge should be evident to all of the parties involved. The final decision should not be a surprise to anyone.

If you follow the "go slow and be sure" guideline, then the effect on the facility, the process, and the remaining employees should be minimal. The people in the facility want their coworkers and themselves to be treated in a fair and equitable manner. They realize there are rules and performance expectations for all of the workers in the facility, including management. The perception of how these rules are applied goes a long way toward achieving the desirable atmosphere in the plant.

When an employee performs poorly, determine whether it is a onetime occurrence or an ongoing problem. Everyone makes mistakes from time to time. Good employees should not be "raked over the coals" for a rare slipup in performance, but you shouldn't let it go unnoticed either. A quiet, private conversation to review the cause and effect is definitely in order.

PART TWO: KNOWING THE PEOPLE

In cases in which an employee's poor performance is chronic, your first step is to counsel and retrain. You have spent a fair amount of time and money getting the employee to a certain level of knowledge. Maybe some remedial training in a weak area is all that is required to resolve the problem.

If the employee's performance doesn't improve, then follow the required steps to remove the employee from that position. Document your comments, findings, and efforts in a log with dates, times, and what was said. Have witnesses present from management and the work force, union or nonunion, if a discussion takes place with the employee. It is very important to emphasize how the person's performance is affecting the process. The company expects all of its employees to manage the process.

An important point to remember is that the rest of the employees in the process know who the poor performers are in the group. They expect management to know who these people are also, and they expect management to do something about it.

Employees feel this way for two reasons: First, they do not want to be responsible for another person's poor performance. It usually means they will have to work harder at their own jobs. Second, they want the process to run well, because it makes their jobs easier. More importantly, a well-run process provides the sense of job security we all value.

It can be difficult to identify who is not pulling his or her fair share of the load. The more you know about your process, the easier it will be for you to pinpoint the human bottlenecks.

Collect the data you need to illustrate the poor performance. Look at the number of quality defects the person is generating, their average output compared with that of similar operations, and the downtime totals. Remember to go slow and be sure; you are talking about challenging a person's livelihood.

Don't make the mistake of confronting the group if the problem involves just one person in the group. Don't be like the teacher who keeps the whole class after school because one student misbehaved. People don't like this method, whether they are children or adults. In either case, the students

THE SOFT ISSUES

or the process group can create more problems when they feel they have been treated unfairly.

Finally, make sure you or someone in your facility (the personnel manager) knows the contract language on termination backward and forward. If a situation arises in which a suspension or termination is the possible result, then the steps must be followed to the letter for the protection of the facility and company as well as the employee. If no one is sure about the termination process, ask for help from your corporate office.

Quite often, you will know the language better than the employee and union steward, but don't assume anything because of your superior position. The termination process challenges your facility's process and the employee's means of supporting his or her family. *Take it very seriously!*

MEETINGS

The thought of attending a meeting conjures up images of wasted time. Most of the experiences people have had with meetings justify these beliefs. Managing a process environment requires that you and your people use time effectively. Wasted time on any task means less time devoted to improving the process. Wasted time in meetings is a time management sin for a process manager. There are some things you can do to make meetings more productive.

The first thing you should do is to ask yourself if you have to have a meeting. Can you address the question with a memo to and short conversations with the appropriate people? Always choose this latter method if you can.

I ran a process environment for 3 years without having one formal meeting with the line supervisors. I accomplished this feat by writing a lot of short memos and having a lot of short, informal meetings each day. In all fairness, the process parameters were reasonably established. The supervisors had been at their jobs for quite some time and we did not have to deal with any new, big projects during that period. It can be done, but it requires more effort from you. You have to be on

PART TWO: KNOWING THE PEOPLE

the floor and involved to keep up with each of them and their activities.

The second thing you need to recognize is that there are three kinds of meetings: "me"-tings, "meet"-ings, and "meat"-ings. You have probably attended all three kinds at one point or another in your career. Understanding the differences between the types of meetings can help you with the productivity of your meetings.

The first and least productive kind of meeting is the "me"-ting. This type of meeting is usually called by a manager who is not involved in the flow of communication in and around the process. This meeting is usually scheduled for the same time and same place each week. A meeting of this type is characterized by no agenda, no minutes, and unfocused discussions. The people attending the "me"-ting are required to listen to what people have to say as each person takes their turn at playing "Mr. Microphone." Ninety-five percent of the information in these "me"-tings can usually be handled with a memo and/or some individual conversations. The format is "let me tell you what I am doing."

The second type of meeting is the "meet"-ing. This type of meeting is usually for information dispersion and/or gathering on a particular topic. There may be an agenda, but there are seldom minutes. Many times, there are visitors from corporate—this is usually the reason for the "meet"-ing. Depending on the culture of the company or the process, a free exchange of ideas and concepts on the topic may not be encouraged. The format allows people to meet each other, but time is usually not available to explore and develop ideas.

The last type of meeting is the "meat"-ing. This gathering is the most productive, because it has substance. It has something you can sink your teeth into; it has "meat." It is characterized by its organization and its atmosphere. There is always an agenda or source document to control the discussion. In the best of worlds, the agenda has been distributed the day before, so each person can read it and be prepared.

The discussion is open, honest, and freewheeling. The attendees are not pressured by politics to be quiet or to carefully avoid saying the wrong thing. The only wrong idea is the

THE SOFT ISSUES

one never said. The person in charge of the meeting keeps the discussion moving and redirects the follow-up questions when necessary.

Avoid trying to accomplish too much in one "meat"-ing. If there are four really important ideas to discuss, but you think there isn't enough time, have two shorter meetings. Most people can't focus effectively for more than 1 hour on any one topic.

A sure sign that a meeting is going on too long is that people are leaving to answer phones, get coffee, or go to the washroom. An organized, well-paced meeting has to be quite important to keep everyone away from the process for more than 1 hour.

Request minutes or follow-up documents with names and dates assigned to specific projects for all meetings. The minutes must be in your "in basket" no later than 24 hours after the meeting. If the person who called the meeting is organized and ran a good meeting, he or she should have been able to record the minutes during the meeting. Let the typed copy, if it is necessary, follow later. If the document is going to stay in the facility and the handwriting is reasonably legible, then keep the secretaries busy with more productive tasks than typing.

People get used to and can even read the worst handwritten notes. Handwritten notes are also more personal and not as threatening as typed notes. They can even be a source of humor as people try to decipher the handwriting.

Impromptu meetings can be great fun and productive brainstorming sessions. Yet, it is unlikely that you or your managers' schedules will accommodate this format very often. In most cases, an agenda should be issued. Encourage your people to generate an agenda and issue it the day before the meeting. It forces them to plan and think out the meeting process in advance.

Who should attend the meeting? Anyone who can or should contribute to the topics being discussed. Yet, too many people at a meeting can be difficult to manage. Each person controlling a meeting has a different skill level for controlling the flow in a meeting and getting the participants involved. Most people can control five to eight people in one group; others can make 20 people feel involved.

PART TWO: KNOWING THE PEOPLE

A meeting can also be an effective development tool for line managers and line operators. You should consider inviting a line operator to make him or her feel more involved. Everyone has a natural curiosity about what is happening in meetings they are not invited to attend. Rotating different people into the meetings will help reduce the rumors that can develop from this curiosity.

The meeting room should have a variety of visual aids to help the people present their ideas. Chalkboards, video equipment, overheads, and on-line computer-generated overheads can be utilized.

There is software available for "computer meetings." This software lets the group, as individuals, anonymously manage the agenda and the ideas discussed by showing all of the comments on a large overhead display. It increases productivity, because it reduces the fear some people have of saying the wrong thing and being embarrassed. I guess this could be called a "machine-too" or a "(me)-too" meeting.

A good place to start to improve the productivity of the meetings in your process environment would be with those "same time, same place" meetings. Make them every other week or once a month. See if it makes a difference. Ask people to get together in smaller groups if they feel the need. Maybe two or three regular meetings can be consolidated into one regular meeting.

You will know you are on the right track when you hear someone say "That was a good meeting." It can be done, but you have to work at it.

SAFETY

Safety is an issue that deserves as much of your time as anything else. Many managers feel lost when it comes to managing safety. You will hear the comment that avoiding an injury in a process environment is "mostly luck." The fact is, you can have an impact on the safety record at the facility. Some of the approaches are direct and some are indirect; together, they can produce good results.

THE SOFT ISSUES

The first thing you have to understand is that the money and time you invest in achieving a safe environment are worthwhile. Accidents cost the company money. In the short run, you lose a trained employee, and the replacement employee may not be as knowledgeable about the process. Quality and efficiency may slide backward. There is also the extra cost to the system to hire or transfer an employee to that position.

In the long run, the insurance and workers' compensation costs will increase with each accident in your facility. These costs will stay with you for 2 to 3 years, even after you have made changes to correct the situation. The insurance companies make money on the probability of injury. By keeping a 3-year average, the insurance company can charge a higher premium for a longer time. Therefore, you must manage the process for a consistently safe performance over a long time to achieve a low probability of injury, which means lower premiums.

On the human side, no injury is pleasant. Some injuries heal quickly, but others never go away. Some people may become disfigured or handicapped and, make no mistake about it, some people still die. There is enough electricity around any machine to cause a fibrillation of the heart muscle. When an employee is injured, the feeling and memory stay with him or her forever. I have had two experiences that still give me a weak-in-the-knees feeling when I think about them.

The feeling and memory areas of the safety program are the indirect issues you need to address. The best safety program starts with the correct mindset. All of the employees should have the right psychological approach to the environment, and all of the people working in the process have to "think safety." Their thoughts and actions to promote a safe environment have to be valued and periodically rewarded by their peers and management.

Developing the correct mindset starts with the person at the top. The people have to believe that this person cares more about their safety than anything in the process environment. They need to hear that commitment in group and individual situations. Safety meetings involving only safety committee members cannot carry the message to everyone. Notes, bulletin boards, visual aids, posters, and just plain one-on-one talks

PART TWO: KNOWING THE PEOPLE

should be part of the program. You and the rest of the management group have to believe!

There are several ways to directly affect the safety of the process environment. One part of the program is to investigate the cause and effect of every accident. Determine how and why it happened. How could it have been avoided? What steps need to be taken to prevent its recurrence? Maintain a detailed log of all accidents. The investigation should be treated as a process within the process. Keep improving your technique to the point of anticipating accidents in similar but different areas of the process environment.

The safety committee should review all investigations and the required follow-up procedures. The committee should include representatives from the management and line operator groups from each shift. The members should collect suggestions from other people in the process. As with any meeting, the safety committee should issue an agenda and detailed minutes.

You need to recognize that investigations, follow-up procedures, and meetings take time. The people assigned to perform this function will need some special considerations or relief from their regular duties. Timeliness is important. Accident investigations and follow-up procedures have to be quick to be effective.

You may have to deal with an employee who "creates" an accident for his or her benefit. Such situations are rare and very difficult to prove. Back injuries are probably the most common and difficult to diagnose. Everyone is aware of this difficulty, and a few people will take advantage of it.

You can do several things to help minimize this problem. Screen new employees for this type of injury by asking on the job application for any history of this kind of problem. Ask the employees to wear back braces while they are at work. As you learn the personalities of the different operators, try to determine which ones may try to have an accident. More importantly, identify which employees are most likely to have an accident because of poor work habits. Retrain or reassign these operators. Look at every machine and work practice as a potential accident in the hazard analysis of your process.

THE SOFT ISSUES

The safety committee members will usually need the resources of at least one other department to perform their jobs. Make sure the other departments are there to provide the required support.

A common pitfall for safety committees is the regularly scheduled "meet"-ing. People in this situation lose contact with the true safety process; get the people out of this pattern. Have them spend the meeting time analyzing one area of the process. Bring in specialized training presentations on first aid, spill control, cardiopulmonary resuscitation, and emergency evacuation procedures.

Have the committee visit the facility on the weekend to observe the shutdown, sanitation, maintenance, and start-up processes. Have them examine the different pieces of support equipment, including ladders, walkways, and carts. Maintenance should replace ladders, for example, but "new eyes" from the safety committee can sometimes see things that others who are there every day cannot.

The committee should be involved in the preparation of the training manuals for repairing and cleaning the equipment. An effective training manual must cover the safety issues of that part of the process. It helps to observe the machine being taken apart and put back together to understand what is dangerous about the task.

Finally, safety is everyone's job. People have to look out for each other. Management has to recognize that a safe act may take a little longer to perform. It may require some different equipment or more preventive maintenance. Whatever it takes, you need to support safe thinking and procedures to take the luck out of the safety program.

CHAPTER 5
THE DEPARTMENTAL PROCESSES

THE LINE SUPERVISOR

If any one person actually "owns" the line, it is the line supervisor. He or she has to channel all of the resources into the line and achieve the result: a high-quality, efficiently produced product. The line supervisor brings together the efforts of the individual workstations contributing to the process. The line supervisor has to train and motivate the people, control the flow of materials to the process, and monitor and direct the technical aspects of the process. He or she has the people on the line to help perform the task, but the supervisor is always there to maintain harmony among all of the resources.

Managing the line supervisors is not an easy task. Many process managers are too consumed by the perception of their "true" responsibilities to devote the appropriate amount of time to managing this very important group of people. If a manager perceives that his or her role as the line supervisors' boss exempts him or her from helping the process improve each day and helping the line supervisor achieve the improvement, then he or she has the wrong perception.

PART TWO: KNOWING THE PEOPLE

Improving the process is everyone's job every day. You achieve this goal by helping the line supervisor be a better supervisor each day, week, and year. You have to develop the skills of the people around you.

I had a supervisor who was frustrated with a manager's performance and commented, "I pay him enough money so I don't have to hold his hand." My supervisor had a great many skills, but recognizing the weaker skill areas in his subordinates and developing each of those areas was not one of his strengths. Money was supposed to take care of development.

A fundamental truth of managing anyone is that it is okay to hold a subordinate's hand. In fact, it is your duty to hold his or her hand when necessary. Everyone can always learn something from someone else. You never stop learning, and learning to be a good supervisor is not easy for most people.

The least efficient method of learning is by trial and error. In a process environment, learning everything by trial and error would take too long and be very costly. Teaching the supervisors how to manage their parts of the process is much more efficient and cost-effective.

So, how do you develop the skills of the line supervisor? The developing process starts with you. Look at your teaching skills. Can you explain things clearly to people of different abilities? Are you a good listener? Do you hear what people are really trying to tell you? Do you recognize the strengths and weaknesses of each individual supervisor? Can you get people excited about their responsibilities? Can you get people excited about the process?

Try to remember the skills of a good teacher you had during your educational experiences. What made him or her a good teacher? The same skills that worked wonders for you in the classroom will work wonders for you in your process. You have to explain, listen, and motivate effectively. You have to teach the supervisors to listen to and motivate their people.

The most important skill of the process teacher is to recognize the individuality of each supervisor. You have to adapt your explaining, listening, and motivating skills to match the individual needs of each supervisor. The supervisor, in turn, should react to each of his or her people in the same manner.

THE DEPARTMENTAL PROCESSES

How do you recognize individual differences? Again, start with yourself. Identify your strengths and weaknesses and compare them to those of each of your subordinates. Make a chart to help you visualize the differences. An MPPY rule of managing is as follows: *Develop the strengths, and the weaknesses will improve.* Everyone cannot become a superstar. Each person has the ability to continually improve, but that doesn't mean everyone will improve to the same level.

There are many skill areas you can use to evaluate the people in your process:

- *Analytical:* good at figuring out cause and effect

- *Time management:* completes tasks on time and has an organized desk or workstation

- *Problem solving:* can identify a problem and fix it

- *Creative:* can generate a variety of "what if" possibilities for a given question and is good at design and layout

- *Teaching:* explains things clearly, listens carefully, and recognizes individual strengths and weaknesses

- *Honest:* tactfully says what he or she thinks and offers an accurate perception of process events

- *Accessible:* shows their human side to the people

- *Visible:* spends quality time in the process environment

- *Enthusiastic:* is positive about the process and the people

- *Numbers:* good with computers and understands statistics

- *Writing:* can prepare concise, well-conceived documents

PART TWO: KNOWING THE PEOPLE

When you recognize and understand individual differences, you can set specific goals to help each person improve. Communicate these goals to the supervisors. When the supervisors know what is expected of them, they will more effectively communicate these goals to the line operators.

Provide regular feedback on their performance toward achieving the goals. Give positive feedback each time something is executed well. Saying "thank you" and "good job" each day is a simple and effective positive reinforcement tool. Use situations in which improvement is the most likely outcome to create a positive learning experience. Focus on the areas that need work, but don't dwell on them. End conversations with positive, "can do" types of statements. Frame your statements in a positive light: the glass is half full, not half empty.

Here are some other suggestions to help manage your line supervisors more effectively. Keep their paperwork to a minimum. You cannot manage a line from behind a desk. If there are paperwork requirements, computerize as much as possible to increase speed and avoid duplication. If you have a secretary, then maybe the line supervisors should have one for their paperwork.

Be aware of how much time the supervisors are spending in the building. Don't let paperwork and meetings keep the supervisors in the facility more than half an hour longer than the people working for them. You have to keep them charged up, not burned out.

I had a conversation with a process manager about burning out line supervisors. He said, "Yes, we probably push them too hard. Maybe they can last 2 years before they have to give it up. But there will be others to take their place." This manager was incredibly insensitive to the human needs we all have.

The line supervisors in this facility were aware of this attitude and the lack of support for their role. It showed in their efforts. Some of the best people did leave, and those who stayed learned to adapt by not realizing their full potential. They were unhappy, and the result was a process that did not perform to its potential. An MPPY "people" rule is as follows: *Good people will always be the first to leave when they realize there is no support for keeping the process in control and improving it.*

THE DEPARTMENTAL PROCESSES

Above all, don't encourage "facetime" in the culture you are developing in the process environment. Facetime is staying in the facility longer than is necessary to give the impression you are working extra hard. Instead, reinforce good time-management skills. People need a life outside of the company. The supervisor should never be forced to neglect family obligations to keep his or her job. Yes, sometimes situations will require extra attention, but it should be the exception rather than the rule.

The line supervisors should be involved in quality review discussions. They need to understand what is good, what isn't, and how their part of the process is doing. They should also be involved in discussions of the competitors' products.

The line supervisors should always be given the opportunity for input on a project being brought to the facility. Ask them for their ideas on the specifications, the layout, and the equipment. Keep them informed of events that may affect the process they are managing during the installation and start-up phases of the project.

Is there a magic number for a supervisor-to-line operator ratio? We expect the teachers of our grade school children to manage the classroom processes at ratios of 30:1. This ratio is even higher in high school and colleges, where the students are older and more mature. The right number for your process depends on the duties you expect the line supervisor to complete. If the paperwork requirements are increasing the supervisor-to-line operator ratio, then there is too much paperwork.

The layout of the process will also affect the level of supervision. Is it on two floors? Are there separate buildings? Is the process spread over a large area?

The most important factor for determining how many supervisors you need or think you need is how far along you are in understanding the science of the process. If the process is really under control, then you will need less supervision. Every line operator should know what to do.

In my grade school, there were up to 70 children in one classroom at the same time. It worked because the nuns had complete control and knowledge of the process in a small area.

Managing a multiple shift operation makes communicating effectively with the line supervisors more difficult and more

PART TWO: KNOWING THE PEOPLE

important. Personal communication with each supervisor every day may not be possible. In that situation, leave notes.

Control of the process is transferred to the second- and third-shift supervisors and line operators. In this situation, trust and knowledge of the science of the process are critical to the performance of the process. These people are managing the process for the company and for you in your absence.

One of the most challenging personnel issues you will encounter is finding the right people to manage the process and the facility on the off-shifts. The best people do not want to make a career of working from 11 PM to 7 AM. The same people principle is true for the line operators. As positions become available on the day shift, people will move up to fill the openings.

A critical part of any 24-hour process is maintaining consistency across all three shifts. When supervisors or line operators are promoted to the day shift, the performance of the process may deteriorate until their replacements are comfortable with the process. Good training and communication programs will minimize the loss of knowledge and experience when people are promoted.

The science of the process or the business needs, or both, determine how the people are scheduled to work. The best a process manager can do is to develop a schedule that is the least disruptive and most pleasant to the people. The best schedule will translate into the best opportunity for achieving the best process.

A common problem in multiple-shift operations is teamwork. Each shift and manager has a desire to be the best. Competition naturally develops between shifts or similar processes on the same shift out of the pride each group has for its role in the process. Each situation has to be carefully monitored so the groups work together and the entire process remains productive.

Sometimes, however, the shift with the best numbers may not be the most productive shift. In many facilities, the first shift has a larger maintenance staff. Major repairs are often nursed along until the first shift is there to repair it. In this case, the productivity of the first shift is reduced.

There may be an "expert" mechanic/operator on the first shift who is the only one who really understands the effect of

THE DEPARTMENTAL PROCESSES

the process on the machine. This person should be temporarily assigned to train people on the other shifts. Identify the experts in your process and make sure their knowledge is spread around to the other people and other shifts. Record their special skills on videotape and in the training manuals.

This kind of delayed repair is fine if the process is set up that way, but repairs intentionally delayed to make one shift look better than another cannot be tolerated. Competition is fine when it is healthy, but the lack of teamwork between the shifts will only lead to a deterioration of the process. The managers and operators involved must be confronted immediately.

If you are new to the process, you may find yourself walking into a long-standing feud between shifts. You should try to work the people through the problem, if you have time. If the situation is more desperate, and improvement is not coming easily, consider reassigning the managers to each other's shift. In a worst-case scenario, you may have to remove a manager from the process.

Simple things such as making sure the exiting shift leaves the work area clean and organized for the entering shift can be a problem for shift supervisors who are not communicating effectively with each other. The best way to manage the changeover is to expect more than a clean and organized area from the exiting shift. The "baton should be passed on the run," as it is in a good relay race. The exiting shift should leave the entering shift with at least 1 hour's worth of supplies. This setup leaves the new shift with enough supplies to deal effectively with other beginning-of-shift issues, such as labor assignments, product changeovers, and equipment repairs. It also forces the exiting manager to think ahead.

At the changeover, each manager and operator should spend a few minutes discussing what happened during the shift and how the process is performing. If a log is kept (there should be one log and not two), it should be in a location each supervisor must visit. A computer database log would be best, but a large notebook will also work.

The comments should be brief and to the point. They should describe cause, effect, and action taken. The supervisors and line operators on the shift should not feel like baby-sitters

PART TWO: KNOWING THE PEOPLE

taking care of the process until the next crew gets there. You want it running well when you are there, when you are gone, and when you come back!

Spend time on each shift getting to know the people and their efforts. It may be difficult to be there at 3 AM, but the regular line operators have to do it every night. Your appearance at this time will go a long way toward creating a truly accessible image.

Have a private discussion about where they are as a supervisor and as a person every 6 months. It can be a meeting to just let him or her talk, as he or she may just need to vent feelings.

Finally, ask the supervisors about your performance. Are you helping them perform their job? Are you hindering their performance? What else can you do for them? Don't let your ego or your position get in the way of learning something about your style. You can help them improve their skills by letting them help you improve your skills.

THE QUALITY CONTROL DEPARTMENT

The topics of quality standards and how the people affect the quality program were presented in the previous sections. This section focuses on quality control as a department in the process.

The quality control department should provide technical support to the process. The quality group should provide consistency and organization to the current process parameters and serve as a guide to the explorations into new processing ideas.

Creating the mindset in the quality personnel that allows you to achieve this goal will be a significant challenge. Quality personnel, by the nature of their training, have a tendency to develop "doomsday" attitudes toward the process environment and the product. They see all of the things that can go wrong with the process, and when they look at the product, they see its faults; the glass is half empty for these people.

People working in the process must see the glass as half full because there really are many, many things that can go

THE DEPARTMENTAL PROCESSES

wrong. If they focused only on what could go wrong, they would never make it through the day, the week, or the year. Their effectiveness would be diminished as their ability to act became paralyzed by the possibility of something going wrong.

The line operators have to maintain the big-picture perspective to realize their daily goal of making the process and the product a little better; the glass has to be half full for these individuals. You should consider this attitude an important personality variable when hiring quality personnel, line operators, and line supervisors.

The uneasy tension between production and quality personnel stems from the different philosophical (half full vs. half empty) perspectives. The tension in some cases can be so great that quality people avoid going into the process areas. Instead, they use the comfort and safety of the lab to pick apart the process and product. They are driven there because the demand for low-cost production is stronger than the demand for quality production and/or because the production people outrank them in the management hierarchy.

Your goal is to create a perspective where the glass is always half full for both groups. They need to completely understand the limitations of the process parameters that would cause them to see the glass as half empty but have the confidence in that understanding to lead them back to the half full glass.

An MPPY rule for control is as follows: *You must control the process; do not let the process control you or your attitudes toward the process and each other.* Let the things that work correctly for you in the process drive your understanding and improvement of the process. Don't allow yourself and your people to dwell on the inefficient and negative aspects of the process. You want the quality and process groups to be the explorers and watchdogs of the process environment simultaneously.

Manufacturing executives have believed that the pressure for lower costs limited their ability to be completely committed to quality. This corporate philosophy dominated operations management almost from the start of the first assembly line until recently. Understanding the limits and science of the process

PART TWO: KNOWING THE PEOPLE

will provide the information on the degree of quality that can be achieved and the related cost for achieving this level of quality. If the limits are acceptable, it is the responsibility of management to get the process there. If the limits are not acceptable, then change the limits or the process or get out of that business.

It may be that the process can achieve the desired quality level by its design, but the knowledge level of the process science is not there yet. The statistical data on the process will help answer this question.

The distinction between these two points is fundamental to managing any process. You cannot be sure design flaws are holding back the performance of the process until you have achieved the level of understanding about the process that proves it to you and to the rest of the line operators. It is a difficult challenge.

Many process managers will say the design is limiting the quality and performance of the process because it is the easy way out. Buying new equipment to fix the problem is easier than analyzing, experimenting, and troubleshooting your way through a problem.

Here are several ideas you can use to bring the line operators and the quality control groups together. Have the line operators go through a quality scoring review with the quality people regularly. The process people need to understand how each score is determined and its importance in the overall score. What part of the process caused a score to be good or bad? What can be done to improve that part of the process?

Make sure there is agreement on what is a quality product. Invite the marketing and sales people to participate in a quality review. Discuss with them the improvements or changes you can make in the product. If they can sell twice as many units with the improvements, then any additional investment, if necessary, will be worth it.

The line operators should come away from the quality review meetings with a better understanding of where to focus their time in the process to improve product quality. With this approach, you may see a request for new equipment or a request to fix equipment that is affecting the quality. Maybe

THE DEPARTMENTAL PROCESSES

new procedures will be developed that improve the process control in the critical areas.

Have the quality control people run the line during the vacations and days off of the regular line supervisors. This actual line experience will help these technical types relate to the "real world." It will also provide a good training experience for future line supervisors. Hire your technical people with the thought that they may be line supervisors at some point in the future. In addition, line experience for the technical people will reduce your supervisory costs, as it may eliminate the need for a relief supervisor.

It may be that the quality personnel and the line operators are spending their time in areas that have little impact on product quality. Are you asking them to spend time in those areas by virtue of the activities you have set out for them to perform or think they have to perform? After running the line, the quality people will have a better understanding of the activities that add value to the process. Make sure that all activities the people perform add value to the quality of the product. Don't let them get buried in paperwork.

Regular group discussions on product quality involving all the people in the process are the cheapest and most effective quality training programs. This type of review should stimulate the development of an interactive, exploratory relationship between the quality, supervisory, and line operator groups. In the end, their shared understanding of the line and their cumulative abilities to improve the process should lead to improved product quality.

Another idea you can use to promote the process-quality relationship is to ask the quality and process people to tour the line each day. Now you say, "But they already are doing that." Yes, I bet they are, but are they taking the tour together? The typical reaction for process people is to head the other way when they see the quality people coming into the process environment. Their reaction is, "What's wrong, now?"

Taking the tour together will do a couple of things for each group. It will reduce tension by improving each person's understanding of the other person's perspective. It should help

PART TWO: KNOWING THE PEOPLE

the process supervisor see all of the things that might be wrong before the quality person sees them. It should help the quality person see the things that are working correctly in the process. The tour should help both groups move toward the "glass is half full" perspective.

Last, but not least, make every effort to keep the quality people out of the lab and in the process environment as much as possible. Keep them involved in the flow of information about the process and use their technical abilities to the fullest. This challenge should include not only analysis but analysis with conclusions and recommendations. The quality group, or any group for that matter, cannot develop sound recommendations to improve the process without being in the environment regularly.

The quality control department should report directly to the process manager responsible for the entire facility. The primary goal of the quality manager should be to maintain and expand the technical knowledge of the process and, most important, to communicate and translate this information to the process manager, the supervisors, and the line operators. They should keep abreast of all of the developments and ideas the people are working on to achieve consistency and minimize duplication.

This discussion assumes you have a quality department, but what if you don't? What if the line operators and the supervisors *are* the quality department? Well, very little changes in the way that quality issues should be managed. You may want to have one quality person available to make sure the checks are being performed correctly and to assist in the training of new operators. This technical person could also be involved in setting up new testing procedures. If you are thinking of converting to a line operator/quality concept, do it in stages and use the quality staff to provide a bridge between the old and the new programs.

THE DEPARTMENTAL PROCESSES

THE SANITATION PROGRAM

The cleanliness and appearance of any process environment is one of the "damned if you do and damned if you don't" issues. If your process environment is dirty, cluttered, and unorganized, it will hurt your image as a manager and will have a negative effect on the people working in the process. Their work efforts will become sloppy and careless. The pride in their job and workplace will be diminished, which will hinder your efforts to improve the process. Sanitation is a very important part of the culture within a facility.

A company will not consider a dirty, unorganized facility as one its "crown jewels." This perception, regardless of how the process is actually running, may affect future investment decisions. A process and the facility that houses it need new investments to exploit the long-term potential of the business unit.

If your process environment is sparkling clean and organized, your supervisor will appreciate your efforts. The facility should be in a condition that would allow your boss to bring a visitor through it at any time. A clean facility will have a positive effect on the morale and attitudes of the process employees.

The tough question is, How much money do you have to spend to keep it clean? Maintaining high standards for sanitation in a facility requires the time of many people, a large variety of cleaning supplies, specialized equipment, organized storage areas, and, most of all, a detailed plan. Your supervisor expects you to keep it clean, but he or she may not be happy with the amount of money it requires. If it is dirty, then he or she will want it cleaned, but not at a great cost.

So, how can you win? There is only one way to win. There is only one way to have and maintain a clean, orderly process environment in a cost-effective manner. The solution is achieved by developing a detailed, organized plan of attack. Sanitation is a war you wage against a process that is constantly creating things to clean. It is a war in which the people working in the environment must participate as allies. They have to be responsible for cleaning their little part of the process. It takes everyone working together all of the time to achieve a high standard of cleanliness.

PART TWO: KNOWING THE PEOPLE

The line operators should be responsible for those minor daily activities that keep the appearance of the process satisfactory, such as sweeping the floor; putting away their tools; organizing their workbenches, desks, or lockers; and making sure the trash is recycled or emptied. These duties should be part of the line operators' daily responsibilities to the extent that they do not hinder their efforts to perform their primary task, which is to run their part of the process efficiently. These tasks should be recorded as items requiring daily maintenance on the master sanitation schedule.

There will be times when the process goes out of control or the regular operator is absent and these tasks are not performed. The problem can be handled by delaying the cleanup until the next day, by having someone stay to complete the tasks, or by having the next shift pick up the slack. If this carryover doesn't happen all of the time to the same shift, then each shift should be able to help the team out when it does occur.

If one shift is continually being left with a mess to clean, then there is a problem that needs to be addressed. Maybe one shift is performing most of the significant maintenance items due to the way the maintenance department is staffed and the next shift is inheriting the cleanup. In this case, schedule some extra sanitation time for that shift to deal with those situations. If there is no maintenance on a particular day, then have a list of weekly or monthly items that can be cleaned.

The ultimate success of any master sanitation program depends on the person in charge of the program. This person must take an ownership position with the program. The person must regard the sanitation process as one that is just as important as any of the other processes in the facility. You must show your support for the sanitation manager and the program if it is going to be a success.

Sanitation is similar to quality in that you cannot go halfway. You have to make the commitment and support it or lose it. A process will become dirtier more quickly than it takes to clean it and keep it clean, because it takes much less effort from everyone to be dirty.

Finding the right person to manage the sanitation program is not easy. The person has to have a working knowledge

THE DEPARTMENTAL PROCESSES

of the cleaning equipment and the cleaning supplies and has to be aware of the safety issues concerning all of the sanitation procedures.

The sanitation leader should have a basic understanding of the operational procedures of the process. The process equipment could be damaged or altered during the cleaning procedure by the cleaning solutions, the equipment, or the sanitation crew. The sanitation leader should use his or her knowledge of the process to prevent damage.

All of these skills are essential to the success of the sanitation manager, yet the most important skill is the ability to teach and motivate the sanitation crew. The crew has to believe that they contribute to the overall success of the process, and the manager has to be able to relay this message effectively to all parts of the sanitation program, especially the hardest, dirtiest, smelliest, and most disgusting tasks in the program.

The final skill the supervisor must have is the ability to organize the program. The sanitation program will always be changing, adapting, and improving as it follows the changes and improvements in the process. New activities will be added to the schedule as more is learned about the process and equipment. The manager will have to fit the changes into the framework of the plan in a cost-effective manner.

How is a plan developed? First, write down every task that is currently being performed. The tasks should be divided into frequency categories, such as each shift, each day, weekly, monthly, quarterly, semiannually, and annually. Next to each task, record the number of hours required to perform it safely and effectively. You may also want to include the type and cost of the materials and the equipment next to the task. This cost information will be useful when budgeting next year's expenses.

Once you have recorded everything that is being done, determine what else should be cleaned. A simple way to approach this potentially overwhelming task is to focus on one area or room. Choose a room that does not have a lot of equipment in it and that maybe doesn't even get too dirty.

Next, analyze each part of the room to the smallest detail you desire. Divide the analysis into the north and south walls,

PART TWO: KNOWING THE PEOPLE

the east and west walls, the floor, and the ceiling. Consider everything on each surface that needs to be cleaned and determine how often and by what method it should be cleaned.

Don't worry about the cleaning frequency being correct the first time. If an item has not been cleaned for a long time or the process environment in that room has changed, then you won't really know the appropriate cleaning frequency. Once it is cleaned, come back and check it regularly to determine when it is ready for another cleaning. When you complete one room in this fashion, the other rooms will be easier, because cleaning efforts and frequencies are very similar for similar tasks.

A different approach to cleaning the process equipment is required. There have been many discussions replete with four-letter words among line operators on a startup when the sanitation crew sprayed too much water, used the wrong cleaning solution, or did not reassemble the machine correctly.

It is very important to decide what cleaning tasks should be performed by the line operators, the sanitation crew, and the maintenance department. Some managers let the sanitation crew break down and reassemble the equipment; others let the sanitation crew break down the equipment and let the line operators or maintenance department reassemble it. The method you choose will depend on the staffing and skill levels of each group.

Once the master schedule is created, daily work sheets can be distributed to the various crews. These sheets can be checklists that help you and the people remember what is to be cleaned and how. Once it is set up, the program will manage itself, but it will take a considerable amount of time to set it up correctly. It will take at least 1 year for you to see all aspects of the cleaning cycle.

Managing and keeping the program current will be much easier if you put it into a database program on a personal computer. The computer can create the daily sheets for the people and summarize your costs and labor data.

You should also consider displaying the program on a color-coded magnetic bulletin board. The board will not contain the same level of detail as the backup checklists, but it will provide a visual aid for you and your people to challenge. It

THE DEPARTMENTAL PROCESSES

would also be a nice way to show off your organizational efforts.

THE FRONT OFFICE

The people in the front office play a very important role in the efficiency of the process. They order the materials, pay the bills, process the payroll, answer employees' questions concerning benefits, answer the phones, type reports, generate the accounting documents, and keep unwanted salespeople away. The tasks they perform keep the line running and the people happy. A happy group of process people translates into a happy process!

The people in the front office should always be courteous and professional in dealing with the line employees. They should not talk in a condescending manner. Very often, line employees are uncomfortable entering the office; it is clean and quiet; the people are usually well dressed; it might be air-conditioned; and the office staff may have some formal education. The office environment can be quite different from the process environment, and it can be intimidating to some process people.

Everyone can benefit by having the office people visit the process area. Perhaps they have a question about a processing standard or about how accurate some data may be. Encourage them to take a walk through the process area to resolve the question with the line operator or supervisor. Their tendency will be to call the supervisor to the office, but the office staff will be more productive, more a part of the team by getting involved in the process.

Someone in the front office should answer the phones for the period during which the salespeople or the corporate office call the facility. There is nothing more frustrating for an evening shift supervisor than to have to answer the phone when he or she is trying to resolve processing issues.

Keep the office area clean, organized, cheery, and well illuminated. A shared coffee pot can be a simple convenience item worth its weight in gold for the conversations it can generate. Another simple gesture worth its weight in gold is food!

PART TWO: KNOWING THE PEOPLE

Donuts, fruit, or snacks will generate friendly conversations that will relax everyone. People work better in a relaxed atmosphere.

How can you tell if your office group is productive? Are there too many people? Are the right people performing the tasks in the right order? Office productivity is hard to measure, but here are some things that will help you understand their activities.

Generating the payroll is a time-consuming task that has the same deadline each week. If the headcount in the process fluctuates, find the point at which the payroll person needs more help to get the job done. When help is required, where does it come from? If other members of the office staff help, does this delay any other activity, such as generating the performance report?

The headcount and payroll responsibilities may increase to the point where the whole office process comes to halt. When this happens, consider hiring a part-time person to help with the payroll only. There are also personal computer programs that eliminate the time spent to manually calculate the time cards.

Do you pay your own bills? Ask the accounts payable person to keep a list of the bills that are processed. If your business is seasonal, this number could fluctuate, and help may be necessary during peak periods. Determine how many bills can be paid each week by the current staff. In an emergency, instruct them to pay the bills that must be paid to keep the process running; the other bills could be paid the following week.

The accounts payable function is directly related to the issuing of purchase orders. The preparation and distribution of purchase orders also takes time. It takes the same amount of time to prepare and pay a $50 order as it does a $500 order. Payroll and month-end closing days are not good days to prepare purchase orders or pay bills.

When is the daily efficiency report on your desk? Is the process group providing the complete and correct information to the front office? The people in the office should not be spending time trying to figure out what the process report says.

THE DEPARTMENTAL PROCESSES

Make sure the operations supervisors are completing the reports correctly.

How much time is spent supplying information for corporate requests and reports? There may be deadlines for some of this information that delay the preparation of the process reports. If the corporate requests are bottlenecks in your office process, don't hesitate to raise the issue with the corporate office. A corporate office cannot exist without a process to make quality products for sale.

When you ask for a special report or analysis, how long does it take for you to get it? Is there someone creating special reports without being asked to create them? In either case, you will get an idea of how much free time is available.

If someone asks the "what if" kinds of questions about the data he or she is looking at each day and generates a report on his or her thoughts, reward him or her by reviewing the report and responding to it with questions for further explanation. Try to provide more time for this person to think of other things to analyze. Do everything you can to encourage exploration in all departments.

Have a meeting with the office staff about their daily, weekly, monthly, and annual duties. Have the group focus on different ways of performing the tasks or shifting some of the work loads. Some people will happily switch activities that are boring to them with someone else's activity.

Phone calls also take time. People calling about purchasing materials will generally make fewer phone calls than shipping people, but the calls will probably be longer as price, quantity, and delivery issues are negotiated.

Finally, if you have room and money, partitions between the desks are helpful. They reduce the noise level and provide some privacy. Remember, these people generally don't go more than 10 feet from their chairs. A space to help them define their individuality can go a long way toward helping them like where they are sitting.

PART TWO: KNOWING THE PEOPLE

THE RECEIVING AND SHIPPING DEPARTMENTS

The receiving and shipping processes usually report to the same person and are usually handled in the same dock area. These areas operate with systems characterized by detailed paperwork requirements. They are also subject to the outside influences of other companies' processing and distribution environments, which can be very difficult to control.

Receiving procedures start with the receiving document. The receiving document should contain all of the information accounting needs to pay the bills and that the quality department needs for recall purposes. This information should include the quantity received and rejected, the lot numbers, code dates, receiving date, item description, item code, time, and a signature. Checks on the accuracy of the counts or weights are important. The receiving department should verify this information against the information on the purchase order.

Have the receiver move the materials to the correct location immediately to avoid double handling. Maintaining an organized materials storage area is important. The receiver should position the new materials "behind" the older stock so they are consumed last. Consider redesigning the flow and storage pattern in your storage area if this rotation cannot be easily achieved.

Some materials may require sampling and even testing before they are brought into the building. If the test procedure does not require an engineering or chemistry degree, maybe the receiver can be trained to perform the test. Move the test equipment into the receiving area. Make sure the receivers are trained in the proper sampling procedure and know how to spot contamination.

The accounting/purchasing group should develop a list that summarizes the significant items expected for delivery that week. With a little more effort, the list can identify the time the materials are expected to arrive at the facility. If the arrival times are known, you can schedule the labor between receiving and shipping more efficiently. If you order from the same suppliers most of the time and your process consumes a particular material at a constant rate, then the timing of the deliveries can be

THE DEPARTMENTAL PROCESSES

consistent which may help in planning the labor schedule and allocating the space in the warehouse.

An ideal receiving system would be paperless. The receiving information could be entered into a database and compared with the purchasing data that would already be in the database. With a computer terminal at the point of use in the process, the process operator could enter the batch number or assembly consuming the material. The batch number, the code date, and the lot number would be in one place for recall purposes.

If the receiving department is at another location in the plant, consider installing an outside phone/paging system by the door. The driver could page for assistance upon arrival. You can avoid having the receiver "wait" for incoming shipments and can use his or her help elsewhere in the process. If you have more than one receiving area, a computerized system is a must to keep track of the items received efficiently.

Shipping an order effectively has become just as important strategically as making it. Delivering the right order, the right quantity, and at the right time to the customer is a very costly process. The pressure to keep inventories at a minimum and provide the quickest turnaround time on an order squeezes the shipping process from both ends.

This situation will not change. New businesses will do almost anything to get and keep a customer. Mature businesses may be able to pick and choose a little, but they have to be aware that being too choosy will open the door for a competitor to increase its sales.

Different programs designed to "train" the customer on how to order can be implemented to help smooth the processing and shipping schedules. Your company can offer quantity discounts on full-load shipments, price reductions for placing orders 7 days in advance, incentives to receive orders on certain "low-volume" days or during "low-volume" hours, and incentives for certain packaging or loading configurations. Any of these programs, if accepted, will help improve the efficiency of the shipping department.

Unfortunately, most of these programs do not effectively "train" the customer. The reality of the situation is that the customers are under the same pressure to control the costs of their

PART TWO: KNOWING THE PEOPLE

processes. They want you to hold their inventory. Ideally, they would like to turn the product out for sale before they receive the bill from you for the materials.

If you are constrained by a particular common carrier, then shop around. Many companies with transportation equipment are looking for work. Look for a carrier that is willing to stage equipment on your property. You can load trailers and move them out, even if the over-the-road tractor is not there yet. Your responsibility is to keep the finished goods moving into the trailers.

If you cannot stage trailers, maybe you can stage orders in the aisles or dock area. There will be some double handling with this method. If you can only ship on one shift or have some conflicts with the availability of labor or loading time, prestaging some of the loads may help keep things moving. If some parts of an order take longer to set up, then prestaging the difficult portion may help reduce the critical loading path for the rest of the order.

The accuracy of the shipping counts should be spot checked regularly. Errors in shipments to the customer can be very costly to the business. The customer has no incentive to report an overshipment to you. The customer could also claim the shipment was short and benefit from a credit to the bill.

Look at your monthly inventory reconciliations to see if there is a problem in this area of the process. Is one product disappearing at a higher rate than the others? Make sure inventory counts and procedures are accurate. Make sure the information on the bill of lading is being recorded neatly and accurately. Are you really shipping what the bill of lading says you are shipping? If you think the loading counts are inaccurate, have the loaders prestage a complete order. Check their counts against the counts on the bill of lading.

Some of the new inventory software will track your production and shipping by lot code. This program will be more difficult to manage because it requires excellent documentation; however, it provides a better trail to determine which products are disappearing.

There may be certain customers or drivers who always have "short" shipments. A precount may help you verify

THE DEPARTMENTAL PROCESSES

whether the problem is with you or them. You can also send someone to meet the truck if the destination is not too far away. You can check the accuracy of the count with them. Doing this once may stop the problem, because now they know you are checking counts more carefully.

Analyze the flow and storage patterns in the warehouse, and make sure you are taking advantage of every square foot of the storage area. Develop a "paper doll" layout of the area. You can try out different arrangements very quickly and measure the impact on the storage capacity of the warehouse.

Managing the distribution area cannot be done effectively from the distribution office. The manager needs to be involved and out among the "troops." If the shipping requirements have been completed for the day or if the arrival of a truck has been delayed, what does the manager do with the people? Is the area clean and organized? Have the aisles been swept? Do boundary lines need to be painted? Is the equipment clean? Can the people be used in another part of the process? Have the forklifts been cleaned and maintained? Can you send the people home? The shipping manager has to be very flexible to accommodate the constantly changing demands of the area.

PRODUCT DEVELOPMENT

As a subprocess, product development is the only process that incorporates all of the other subprocesses into its structure. The success a product development program achieves is a barometer of how well the other subprocesses and the people are working together in the process environment. A successful product development program unites the knowledge of the science of the process and interrelationships of the subprocesses into a creative clearinghouse for the organization.

New product development is administered in different ways by large and small companies. In large companies, the ideas are channeled through a corporate group and are prototyped on a small scale in a laboratory. Experimenting with a

PART TWO: KNOWING THE PEOPLE

new idea on a high-volume process line is cost-prohibitive because the line is producing product to fill current orders. In addition, the material waste and labor costs would be too great.

Product improvement and cost reduction tests are generally prototyped in the laboratory in large companies as well. The laboratory setting makes it easier to control changes in the product in an organization that produces the same or similar products at different locations. Centralizing the product improvement efforts keeps the process managers focused on their primary task, which is to produce quality product for a profit efficiently. It provides a controlled mechanism for introducing the approved change into the different process environments without producing dramatic variations in the quality.

In small companies, product development will move to the process line more quickly. These companies do not have the financial resources to build and staff a product development laboratory. The higher cost to test on-line is still much lower than the cost to create a separate development group.

In big corporations and small companies, however, the creative forces and the way they affect the product development process are the same. The MPPY rule for product development is as follows: *Everyone should have the opportunity to contribute to the creative process.*

Ideas for product development, product improvement, and cost reduction do not originate exclusively in the minds of the sales, marketing, and accounting groups. All of the people working in the process environment have the knowledge and insights to suggest new product ideas, procedures, and equipment modifications to improve the process. Some people will have more ideas than others, but everyone is capable of generating an improvement suggestion.

This creative process is a direct result of their efforts to improve the process a little each day, week, and year. The enthusiasm to improve the process is a result of the culture the process manager creates to support the continual improvement process. If the culture and attitudes in the process environment do not support a continuous improvement philosophy, many good ideas will be lost.

THE DEPARTMENTAL PROCESSES

People merely punch in and punch out in negative environments. In these environments, when a new idea is tested in the process, the results may appear to be less than desirable because the people are not committed to making it work. For this reason, the selection of a process environment for the first full-scale test is critical in companies with several similar product lines. The selected process should have an MPPY *"we can do anything"* attitude.

Another important source of new ideas is the customer base consuming your product and similar products. Information is available on the sales volumes of these products in the different regions of the country. This information is collected by the UPC scanners at the stores and from the warehouse shipments of your product. The cost of this data collecting equipment is decreasing, and more and better information on product movement is available all of the time.

A new product for your process may be a "me-too" product or a copy of a competitor's product. This tactic is frequently employed by companies in an attempt to gain or protect market share. It may also be a function of meeting the slightly different demands of customers for this type of product or a function of the old saying that "It is easier for a salesperson to sell what he or she doesn't have."

Customers and the market will always demand changes and improvements to meet their needs. The customer can be an end-user, like the parent observing how their child really liked the product and making a mental note to buy it again. The customer may also be the process environment that received a component from your process environment for this product.

In the second scenario, your ability to measure the success of your product and process environment is a function of how many orders you get from your customer. This is the only way you know how many parents are buying the final product. Make sure you understand how your products are being used in the other processes. Visit your customer's process area and discuss ideas on how to improve your product in the process. Use this knowledge to guide your product improvement efforts.

PART TWO: KNOWING THE PEOPLE

The process environment producing the final product must successfully coordinate the efforts of all of its suppliers into the preparation of their product. They must challenge each supplier to constantly improve their product. Your success is indirect, because it depends on the success of the product improvement efforts of the other process environments supplying the other components. The suppliers are linked together by an invisible network of responsibility. The success of each depends on the others.

A new product idea is always managed in the same way. In big companies, the process is more formal, which may slow it down. In small companies, the development process may move more quickly because the cost of measuring every variable at each stage is too high. The small company will probably take more calculated risks, on presumably good assumptions, about their knowledge of the process. Large and small companies are always trying to be somewhere in between the extremes to avoid missing any opportunities.

The process begins with a complete dissection of the idea and how it will affect each of the subprocesses in the organization. A product development group must consider many variables when evaluating the feasibility of new ideas:

1 Is this a completely new and different product? What are some other new, similar products that could be developed from the same process?

2 Is this a "me-too" product? How closely should it and/or can it match the competition's product? Are there any points of difference that can be exploited? What is the quality-value relationship of the current product in the marketplace?

3 What is the expected sales pattern and volume of the product? Who will buy the product? What will customers do with the product?

4 What will the product look like? Color? Height? Weight? Length? Width? Taste? What are the acceptable

THE DEPARTMENTAL PROCESSES

control ranges for these variables? How will the variations in equipment performance and the quality of the raw materials affect the ability of the process to meet the desired product specifications?

5   Can the product be manufactured on the equipment already in a process environment owned by the company? Can the equipment be modified to accommodate the new idea? If a new process line is required, how much will it cost? Where will it be located? What will it look like? How long will it take to build it? Are there any processes available outside the company to produce the product temporarily?

6   What sources are available for the raw and package material components? Are these reliable sources? Are there any cost or capacity limitations on these resources? Are data available on the variations in the product quality?

7   How many units per hour will the process produce? How many people will it require at each of the subprocesses? What duties will these people perform at each subprocess?

8   Where are the customers located? How will the product be sold and distributed?

9   What price will the customers pay for the product? What is the estimated cost to produce the product?

10   What will it take to test the idea? How soon can it be tested? Where should the test be conducted? How much will the test cost?

11   Who will evaluate the test products? Who will determine the next step in the development of the product idea?

PART TWO: KNOWING THE PEOPLE

Considering all of these issues requires a considerable amount of time and energy that should be coordinated by one person. Many people will be involved, but someone must be able to devote time to managing the entire project.

The answers to the process environment questions must be provided by the people in the process environment. The process issues must be considered in light of their impact on each of the subprocesses. This analysis is accomplished by creating a flowchart for the entire process. It is similar to creating a CPM (Critical Path Method) chart for the process. This kind of chart should provide information about the capacity, speed, energy, environmental, people, maintenance, and other issues associated with each subprocess. The current control limits from the subprocesses can be used to determine how well it is performing to the target. Will these limits be acceptable for the new product?

The process chart and the product development questions can also be used to compare different process facilities. When you visit another process or review a competitor's products, create a flowchart of their process. These charts can provide a benchmark for your process and product and where to improve them.

Not all of the questions need to be completely resolved before the test begins, but if you discuss the issues, you should identify anything that might stop the project before it begins.

When the test begins, record how well it performed at each stage of the process. When the test is over, compare the product to all of the original assumptions. Determine what the next step should be. Retest? Modify the assumptions and retest? Modify the equipment? Buy new equipment? Stop the project?

The process of developing new products and/or implementing product improvement ideas takes time and costs money. The more complicated and expensive the product, the more time and money it will take.

The product development and improvement processes are vital parts of any company. Survival in the marketplace is contingent on developing new products and methods. There is no product, process, or company that cannot be challenged by a competitor.

THE DEPARTMENTAL PROCESSES

Finally, it is very important to understand that there are really no failures when you try out a new idea. After the experiment is over, you will still have learned something new about the capabilities of your process. This new knowledge may help improve your current process and products or it may help develop and improve a process and product that has not even been conceived yet!

PART 3
KNOWING YOURSELF

Successfully managing a process environment requires a knowledge of the science of the process, a knowledge of the people in the process, and a complete understanding of yourself. The three MPPY elements are linked together in such a way that performing well in one or two of the areas will not bring the process to its highest level of success. It takes all of the elements working together to realize the full potential of the process.

If any one of the three elements is more critical than the other two, it is the third one—knowing yourself. Many processes not only exist, but also achieve some degree of success by relying on a knowledge of the process and the skills of the people who have been running the process for years.

However, the most successful processes are always characterized by the skills of their leaders. These leaders distinguish themselves and their units by forging the process, the people, and themselves into one entity. The organization becomes an amalgam of its original personality and its leader's personality.

The successful process manager must know where he or she is with his or her skill development and must have a good understanding of his or her strengths and weaknesses and how to exploit them. The manager must understand how these skills affect the skills of the other people and the culture of the process.

The successful manager must know where the science of the process has been and where it is going. He or she must define the goals of the process for the people and must determine the reasonableness of these goals.

The successful manager must understand the strengths and weaknesses of the people involved in the process as well as the culture that has developed around them. The manager must understand how these people have contributed to and continue to support the cultural environment.

The manager must, simultaneously, know where he or she is with his or her skills, where the process is, and where the people are with the process. The balance among these three elements is constantly changing. The ability to make adjustments to each of the three areas and to keep the process in balance is the most difficult task facing the process manager.

The process manager is like the captain of a ship moving toward its port; once in port, the ship will deliver its product. The difference is that the process manager is responsible for improving the ship, the crew, and the product while it is moving forward.

The successful operations managers in this decade and beyond will have the skills of a psychologist, a teacher, an engineer, and an accountant. The primary skill areas will become motivation, organization, troubleshooting, team building, and instruction.

These goals can only be achieved by developing and maintaining the correct perspective of yourself as a part of the process and a part of the team of people running the process. Never allow yourself to be viewed by the people as not being a part of the process and a part of the team. Being a leader and manager implies being there for them and the process.

This faith in you is the cornerstone of the belief that the process has a viable future. The enthusiasm and energy the people must have about their jobs is a prerequisite to continually improve the process and to guarantee its success. This enthusiasm and energy emanate from the belief that you can have a positive effect on the future of the process and on the employees' role in the process.

CHAPTER 6
YOU

WHY ARE YOU THERE?

You are assigned to work in a process environment in Anytown, USA. Why were you placed there by your company? Why has your company invested in the building and equipment? There is only one answer to these questions. The process is there to make a quality product that produces a profit for the company. The profit potential in the short and long term is the reason the company decides to keep the process running and you and your fellow employees working.

How do you accomplish this goal? Why are you there? Your role is to manage the process that makes the product that produces a profit. You are there to take care of the investment in the building and equipment that create the process. You are there to make sure the process continually improves. You did not pick the product or the process, but you have to manage the process environment to produce a high-quality product.

The process dictates every activity. All of the departments within the process environment are there to support the management of the process. Managing the raw materials, the sanitation, the shipping, the quality, the maintenance, the accounting, and the people are subsets of the main activity. Every department and its related activities must remain focused

PART THREE: KNOWING YOURSELF

on the primary activity of managing the process. The MPPY rule for the departmental processes is as follows: *Any activity that is not supporting the process in some way, shape, or form is a wasted activity and should be eliminated. The process must be the center of all activity.*

Draw an organization chart that places managing the process at the top and all of the other departmental processes below it as inputs. The primary activity is managing the process. Try substituting one of the departmental processes for the main goal of managing the process. Can the activities of that department be the logical focus for the activities of all of the other departments? No!

For example, there has been much discussion that managing the quality of the product should be your main goal. Various management groups formed quality circles to improve the process. They asked for employee involvement in fixing the process. They preached that the employees owned the line and empowered them to fix any part of the process that needed to be fixed to improve the quality of the product.

In each case, though, the activities of the employees are always directed back to the process. It is the investment in the process you are asking the people to care for. These "managing the quality" activities are just one way of helping you take care of the company's investment in the process. The quality activities must be supported by the other departments to be successful. Total success is the result of the combined successes of all of the other processes.

How do you know which activities add value to the process? How can you spot those activities that do not add value? Most of the activities add some value to the process and the product, but even common, ordinary, everyday activities can be improved. It might require changing the layout, redesigning a form, buying new equipment, modifying existing equipment, adding or reassigning people, or changing a work procedure.

Identifying the subtle changes in everyday activities can only be achieved by becoming intimate with the process. The process is telling you and the other people involved with the process what it needs for improvement. Are you listening to the story it is telling you? Can you hear it? Can you feel it? If you can't, there are people there who can, and you need to find out who they are by HAVE-ing it.

Teach and encourage people to keep their eyes, ears, and hearts open to the pulse of the process and all of the sub-processes. Reward good thinking and "what if" problem-solving attempts with positive feedback and the tools to make them happen. Nine different ideas may fail to improve part of the process, but the 10th idea may be the "home run" that meets or even exceeds the original goal. Look at the nine failures as learning experiences: You just learned nine new things about the process!

Identifying the activities that the people believe add value but really do not can be even trickier. They may have been performing a procedure a certain way for 10 years. It may be a suboptimal way of performing the task, but they are comfortable with it.

You are there to improve the process, even if it means changing an old habit. In these situations, it is not advisable to force the new method on the people. If there is time, try to have change be their idea. Try to answer questions about how it will really change the process in a scientific manner. Conduct a test or several tests to convince them the change is for the better.

To achieve process improvement, you have to show process improvement. Success instills confidence, confidence generates profitability, and profitability rewards success. You are there to make profits from the process. You get profits from people who are confident and enthusiastic about managing the process.

Managing the process is a big and important responsibility—you should never lose sight of this fact. The company has placed you in charge of the process, and you will be held accountable for its performance. You will need the people in your facility to be very involved in managing and improving the process. They may even have a sense of ownership, but the buck stops with you for the good and not-so-good things that happen to the process.

If something not so good happens, are you going to tell your boss that it should not have happened but that you are

PART THREE: KNOWING YOURSELF

not responsible because the employees manage that area of the process? Your boss has put you there to manage all parts of the process; such a response would not be viewed in a positive manner. The ownership of the process starts with you.

How many times have all of the line operators been replaced when a major problem arises? How many times has the process manager been replaced? If your boss doesn't take care of the problems in his or her area, his or her judgment will be questioned and his or her boss will look at how efficiently he or she is managing the processes. Remember, everyone, including your supervisor, must manage the subprocess they control for the company to produce a profit.

MANAGING AND IMPROVING YOURSELF

An assignment that places you in charge of a process can be an all-consuming and self-consuming task. You will find yourself thinking about the projects, the science of the process, and the people in the process after work and during dinner. At times, you will feel happiness and a sense of accomplishment. At other times, you will feel sadness and a sense of failure. You will even dream about the process. All of these feelings and thoughts contribute to the development of stress.

This level of involvement occurs for every process manager, regardless of how successfully they manage the process. The difference is that the successful process manager's primary focus is on positive ideas, about how to improve the process.

Several factors affect your ability to deal with the responsibility of managing yourself and the process. First, you must consciously accept the fact that problems, opportunities, challenges, surprises, or whatever else you want to call them will occur. Recognizing that surprises will occur is 80 percent of the battle you will face in managing job-related stress. Once you develop this mindset, the problems will become more manageable, resulting in less tension and job-related stress.

An important factor in dealing with stress is your previous experience with the type of process you are managing. If you are fairly familiar with the process, then only a few new "surprises" should surface. These surprises may be more culture related than process related. Try to determine which kind of variable is holding back the process. You can rely on your experiences to get you and the process on the right track.

If you have not had much experience with the process you are managing, remember that processes and people issues are basically the same, regardless of the location and science of the process. As you learn the new process, reflect on the similarities between the different processes with which you have worked. Maintaining the big picture perspective is an essential part of the successful manager's mindset. It reduces the anxiety generated by the day-to-day "bumps" in the learning curve.

Generally, a minimum of 6 months is required to feel very comfortable in a process, but certainly after 1 year you should be more than 80 percent of the way there. You cannot predict when it will happen, but a situation will arise to which you will respond appropriately. Later, when you reflect on how you responded, or perhaps even as you are responding, you will become aware of a calm feeling that results from making the right choice and the people responding to you with respect for your knowledge. Reaching this point is a sure sign that you are understanding the science of the process, the culture of the process, and yourself as a part of the process.

Reaching the point where people look to you and are comfortable with your decisions is a significant milestone in understanding how to manage yourself, the people, and the process. Why? Because now you can begin to anticipate the effect of your decisions before you make them. You can review how to execute the decision most effectively. Should you move in the appropriate direction in three small steps or one large step? What will the people and the process be able to handle? Being able to judge the effect your decisions will have will significantly reduce the amount of stress the decision-making process generates for you, the people, and, therefore, the process.

During this adjustment period, and even after, stressful situations will arise. This stress is caused by the pressure to succeed in your effort to create an efficient process. This kind of self-imposed pressure is generally productive for you and the

PART THREE: KNOWING YOURSELF

process. However, it can be destructive if the manager places his or her personal goals above the goals of the process. When this happens, process decisions become political decisions, and the growth of the process stagnates.

Some of the stress can be created by your supervisor pushing you for quick results. Your supervisor should understand what you are attempting to accomplish and should be willing and able to help, teach, and, if necessary, lead you toward the desired performance level.

Unfortunately, some supervisors during the course of your career will not be able to help you develop your skills. This situation can be very difficult and stressful. Your supervisor may not have the right blend of MPPY skills to enhance your development and the development of the process. Many companies will struggle with this problem in the next decade, as they try to find corporate operations managers with the right mix of MPPY skills to help develop the process managers.

When you do not have a supervisor who can foster your development, you should try to find someone you can talk to about what you are doing. Utilizing someone as a third-party sounding board is an excellent way to relieve stress and to put the issues in their true perspective. The person can be a friend, a family member, or someone you worked with before.

A related type of stress generated from the pressure to succeed is created by how you think your boss perceives your efforts. This type of stress is most common in new assignments until you enter that comfort/confident zone.

In this situation, you can help manage stress by writing down some goals that can reasonably be achieved during the first 6 months of the new assignment. These goals give you and your supervisor the opportunity to set some priorities on which to focus your time. Agreeing on some set of goals takes the mystery out of the "what you think your boss thinks" mindset. Use these goals to guide your day-to-day and week-to-week activities.

Each Friday afternoon, make a list of the things you need to work on in the coming week. This brief exercise provides a quick evaluation of what you accomplished that week and allows you to set priorities for the next week. You can go home for the weekend without having to think about what you have to do on Monday. Let the punchlist on the piece of paper help separate you and your mind from work for the weekend; you cannot think about work all of the time.

The real and imaginary pressure you feel from your boss and the pressure you create about your performance are a product of the desire to succeed. Everyone wants to be successful and to have a positive self-image.

It is the desire to live up to your image of yourself and to sell that image to your supervisor and to the other people in the process that creates stress.

We are all trying to create a positive self-image every day by everything we do. The stress and pressure created by the people and processes around us challenge the reality of this self-image. The daily challenges remind us that we are all from the same mold and that nobody is perfect. The daily challenges highlight the gap between our desired self-image and reality. The challenges provide the direction for us to improve ourselves.

How you react to the challenges is very important. As I said, one way is to find someone with whom you can talk about the pressures. You should also try to find some mechanism of physical exertion for emotional release, such as walking, swimming, cycling, jogging, or gardening. Physical activity is the best way to relieve the stress that is affecting your body physically and mentally. A nonphysical activity, such as reading a book, playing a musical instrument, or listening to music, can also reduce stress.

A very effective way to release stress is through some form of concentrated relaxation, such as meditation or a short nap. You can actually train yourself to calm down by being aware of how anxious you are feeling and getting control of that feeling. Many people could benefit from some training in this area.

When you are at work, it is much more difficult to find a way to release emotional stress. If you start to feel anxious, weak in the knees, or insecure about what to do, then walk away. Find a place where you can reflect on what happened and what needs to be done. Go to the rest room and wash

PART THREE: KNOWING YOURSELF

your face. Jot down a few notes on the possible plans you might employ. Recognize that this feeling is not unusual and that it has happened to many people.

If you cannot remove yourself physically, then remove yourself mentally from the emotions of the situation. Take a step back and get the "big picture." Look at the situation through the eyes of the other people. Get a quick assessment of what they are perceiving and feeling. This step accomplishes two things: It forces you to ignore what you are feeling at that moment, and it usually provides you with some direction for resolving the issue.

Effectively managing yourself and the pressures you feel means recognizing that you cannot do everything yourself. You have to delegate tasks to different people in the process. Some projects can be delegated to some people by writing a note. The same projects might require a short meeting to explain them to other people. When you do delegate a project, keep track of the time it takes to be completed. If it is taking too long to complete, then ask the person if he or she needs help.

Recognize that two people given the same project will usually resolve it in two different ways. They will not approach the task the way you would have approached it. The fact that each person solves problems in slightly different ways and not the way you would should not bother you. Look at the result: If the issue was resolved in a satisfactory manner, then accept it. Try to learn something about the other person's approach to the problem. Examining another person's method may provide insights into how to improve your problem-solving method.

Successfully managing any kind of process requires the skills of a large number of people. Some people have strong analytical skills, some have strong people skills, some have excellent time-management skills, some have strong leadership skills, and some have great creative abilities. Each of the skills is important, but it is the blend of the skills that makes each manager, line operator, and process unique.

Each person is different and yet can make positive contributions to improving the process. How can you get the most out of each person? An MPPY people rule is as follows: *Learn to recognize and develop each person's strengths to overcome his or her weaknesses.*

The first step is to understand your own skills completely. You need to know your strengths and weaknesses. If you can pick your own team, finding people who are strong where you are weak could be helpful. It could also be more difficult to manage, as you are confronted with your weaknesses.

How do you identify your management strengths? Examine your past behaviors. At what tasks you did you excel? They may not have been the tasks you enjoyed, but many times they are. Ask other people to whom you are relatively close—for example, colleagues from work or family members—to assess your strengths and weaknesses. Also, a variety of books, profiles, and seminars are available to help you determine your strengths and weaknesses.

All of this soul-searching can be difficult. It may not be easy to accept the "bad news" with the good news. Yet, as a person and a manager, it can be a wonderful growth experience and can provide valuable insights into how you are perceived by other people. This process should be a part of your effort to constantly improve yourself just as you are constantly trying to improve the process you are managing.

The mix of skills you possess is a product of your experiences and your genetic make-up. The kind of environment in which you were raised had a tremendous impact on your present mix of skills. You tried to develop the skills that were valued and rewarded by your parents and teachers. These efforts may not have always blended together very well as you struggled to develop your own identity.

Other skills were inherited from your mother and father, their mothers and fathers, and so on. You had no control over these attributes. The environmental and genetic processes combined to make you what you are today. While these two very powerful forces shaped you, they also created an equally powerful force that you can use to improve yourself. That force is you! You can and must control the processes that are affecting your growth as a person.

You can improve yourself in many ways, through programs at high schools, junior colleges, colleges, and universities;

PART THREE: KNOWING YOURSELF

seminars; correspondence courses; videos; books; volunteer groups; encounter groups; outbound groups; and long walks in the woods by yourself.

Share any learning experience you can with the people around you. Improving their skills will improve the process. The most important point is to do something. Never allow yourself to stagnate. Always try to improve and grow as a person. There will always be new ideas and skills you can learn from something, from someone, and from yourself.

As you broaden your understanding of your skills and gain work experiences, you will develop a clear idea of where you think your career should go. If you enjoy what you are doing, then look for educational experiences that will enhance these skills and further develop your career. If you are not particularly happy, then use the educational experiences to gain a different set of skills.

Changing careers can be difficult after you develop a certain level of expertise in your field. Even though there are many similarities between different processes, companies are generally reluctant to recognize that the overlap of knowledge exists and that it could be productive.

For the most part, people become specialists in the process with which they are working. This specialization is primarily determined by your first job.

There is some truth to the saying "being in the right place at the right time," but it isn't all luck. Having the right educational and work experiences when opportunity knocks is what distinguishes you from the other candidates. Once you are employed, it is your daily effort to improve your skills that will push your career along with your current employer or the next one.

COMMUNICATING WITH YOUR BOSS

You are in charge of the process. You were selected to manage the process for the company, but managing the process does not mean you have to relate every detail of what is happening at the facility to your boss. There are some good things and some bad things your boss doesn't need to know. It is equivalent to keeping certain things "in the family." Communicating details of every activity is not productive for you or your boss.

What you should or should not tell your boss is difficult to determine. Experience will be your biggest help. The MPPY rule of thumb to help you decide what should be communicated up into the organization is *time*. Is this issue going to be important tomorrow? Next week? Next month? If the answer is *yes* to next week or next month, then you should probably communicate the information to your boss. Events that will be inportant for a short time are generally best kept within the process "family."

Some events should always be communicated upward in the organization to you and your boss and/or the appropriate corporate department. The first event is anything that could cause a product recall.

For example, you discover after the fact that part of the process was out of control. Some product produced during this period has left the facility and is now on its way to or is in the market. In the worst-case scenario, the consumers may have discovered the error. Have all of the records for the product in question assembled and work closely with the recall plan or group at the corporate office.

A second event that should be communicated immediately is a significant product hold for product that is still in the facility or company distribution system. Identify the amount and cost of the product placed on hold. Determine how the schedule to meet the shipment of orders will be affected. Will overtime be required to replace product?

A third event that should be communicated is a major equipment breakdown (when the process or part of the process will be shut down for at least 24 process hours). A major breakdown will usually delay the shipment of an order.

An equipment breakdown in a small business can be very costly, because a small business usually cannot afford redundancy in the process design or spare machines to pick up the slack when a breakdown occurs. The small company's response to this kind of problem is overtime. Sometimes the

PART THREE: KNOWING YOURSELF

customer will have to be asked to accept a delay in product delivery.

A fourth event encompasses the category of significant personnel events. A termination that may be contested, a lost-time accident, strike activity in a union facility, deliberate damage to the process by an employee or employees, and union activity in a nonunion environment are examples of events that should be communicated to your boss.

Lastly, make sure you communicate your successes. It always helps to remind your boss how well you are performing. Definitely communicate changes in the process that improve product quality and/or reduce costs. There is nothing wrong with being your own cheerleader. Just make sure you use the word "we" instead of "I" when communicating your successes. Let your boss know it was a team effort.

It is important when communicating any of these events up into the organization to gather your facts. Make sure you indicate that you have the situation under control. Communicate when and how you will follow up with more information, perhaps by writing a short letter on the subject and sending it to the corporate office.

It is likely that the first conversation about the event will take place on the phone, and what you say may not be as important as how you say it. Your communication can include a description of what you are doing in response to the event, but you should also be aware of the tone of your voice. Your tone should be positive and calm to communicate confidence and control. Negative and exasperated tones may raise questions about your ability to remain cool under fire.

It is essential that you understand and react to your boss' style to communicate effectively your efforts in the process. Does your boss understand the process in your facility? Which positions has he or she held prior to the current one? Does your boss form his or her own opinion from the data presented about the situation? Is your boss easily influenced by other people's opinions? Where does your boss fit into the corporate "pecking order"? How is your department perceived by the company? The answers to these questions will provide you with some insight into the proper way to communicate with your boss.

Finally, do you have a clear idea of what is expected from you? Has your boss provided you with a realistic set of goals? If he or she has not clearly articulated these goals to you, you should initiate the goal-setting process. Once again, consider your boss' style and use this perception to determine how formal or informal the goal-setting process can be.

When you set goals for yourself and the process, be realistic. The goals should have some degree of difficulty and require some extra effort from you, but they should also be achievable.

How is the relationship with your boss affected by your desire to have another position in the process or a broader job description? How is it affected by a desire to relocate to another facility? How is it affected by a desire to leave the company? Will these issues affect the perceived loyalty you have toward your boss and the company?

The answers to these questions will vary from supervisor to supervisor and from company to company. You will have to determine what the correct choice is for you based on the communication patterns you have established with your boss. However, for most supervisors and in most situations, you will benefit from discussing these questions with your boss, except for leaving the company. Your supervisor needs to know that you desire additional challenges, even if it means transferring to another facility. Telling your boss that you are ready for a change reflects loyalty to your boss. It tells him or her you are ready for greater challenges. Your performance will be examined more carefully once you have asked for more challenges.

A discussion with your boss of the question of leaving the company must be handled more carefully. You should have a very good professional relationship with your boss if you are asking for his or her opinion on the matter. If you are doing an excellent job for the company, then it will be easier to discuss the reasons for a change. If you are not interested in another opinion, then don't ask.

There are several other ways a person should look at loyalty in a process environment. The most common interpretation of loyalty refers to the amount of continuous service a person gives to an organization. Most companies recognize an

PART THREE: KNOWING YOURSELF

individual's length of service by giving him or her small gifts, such as pens and watches. Recently, many companies have recognized this loyalty by offering early retirement packages to the more senior people in the organization as a way to reduce costs, since the senior people usually have higher salaries than their younger counterparts.

Early retirement programs discount the real value of a long-term employee, which is the knowledge he or she has acquired about the processes in the company. As people move on to other jobs or retirement, the manager of the process must make sure the knowledge of the exiting employee is transferred to the succeeding employee. This transfer of knowledge can be achieved through formal or informal training programs and can be enhanced by letting the new employee work with the exiting employee for a certain period of time.

For other companies, loyalty takes the form of accepting whatever role the company feels you need to perform at any given time in your career. Many companies feel that capable people should be moved every 2 to 3 years to jobs with greater and greater responsibilities. These moves benefit some people and some companies, but moving for the sake of moving really does not help the person, the company, and especially the process.

It is difficult to develop a sense of loyalty to anything if you expect your involvement to be a short-term affair. It takes time to understand the technical parts of the process, the people, and how you fit into the process environment before the process begins to reap the benefits of your efforts. Process environments require stable and consistent leadership from all of the people involved in the process.

Another aspect of moving that you must consider is the impact on your family. Changing schools, making new friends, and learning about different communities in different parts of the country can be a great experience for the family; however, it can also generate a considerable amount of stress. After several moves, the excitement wears off and moving becomes a real challenge. As the children reach their teenage years, the friendships they have developed and the activities in which they participate carry an intrinsic value that usually exceeds the value of the relocating experience and the increase in your salary.

It is at this point that the definitions of loyalty to the company and loyalty to the family start to diverge. What will be right for you is, obviously, a decision you and your family will have to make. In the end, you have to realize that no matter how much you hear that the company is people-oriented, when it comes right down to the brass tacks, the company will do what it thinks is right for the company. The success of the process is the greater goal to which the company is obliged.

The process is greater than the needs of any one individual. There may come a time after you refuse to move that this concept will be difficult to accept. The best thing you can do is to let it go and focus on the future. Do not let the negative feelings that may develop from how the company treated you consume you mentally and emotionally. It is important to put the event behind you, focus on the positive experiences you gained from it, and concentrate all of your energies on your new task. Be loyal to yourself and to your family.

Finally, loyalty to a process and a company does not mean you have to knowingly cover up or contribute to illegal or unethical activities by the company or your supervisor. I think the expression "We all live downstream" summarizes this position perfectly. At some point, someone will have to suffer the consequences of a process that has gone out of control. It is better to keep the mistake within the process, learn from it, and thereby improve the process control than to knowingly release the product into the "downstream" environment where it can damage the long-term profitability of the process and/or hurt a consumer. We rely on each other, working in our respective processes, to be loyal to and to take care of each other.

CHAPTER 7
YOU IN THE PROCESS

STYLE

Successful managers will utilize any one or a combination of the three basic styles of leadership: authoritarian, democratic, and laissez-faire.

The authoritarian manager controls everything. All decisions cross his or her desk. Human or "soft" issues are not relevant to the process, and opinions are not important.

The democratic manager wants everyone's opinion. The group is responsible when a decision is reached, and consensus is important.

The laissez-faire manager walks away from decisions. This type of manager provides little or no direction on issues. Noninvolvement is the key factor in this style.

In a given situation, a manager may use any style, but there is typically one style or actually a blend of the different styles that dominates his or her actions. Very few managers have a purely authoritarian, purely democratic, or purely laissez-faire style.

How effective are the different styles in managing a process? As you might guess, the laissez-faire style is the least

PART THREE: KNOWING YOURSELF

effective style in achieving results. A manager cannot be uninvolved and cannot avoid making decisions.

The authoritarian manager is not the most effective manager. People working in this environment are not involved or committed to improving the process because they look to the boss to do their thinking for them. Creativity is stifled because the authoritarian manager's ideas and methods are not easily challenged. The military model is an example of the authoritarian style.

The democratic manager is also not the most effective manager. Decisions are delayed until a consensus is reached, which means action for improving the process is delayed. Consensus decisions can temporarily force the right decisions into the background. Learning by trial and error is the slowest way to improve the process. The governmental process is an example of this style.

The most effective style is a blend of the authoritarian and democratic styles. The effective manager walks the fine line between these two styles. He or she injects control and direction when people and the process start to sway off course. A nondecision is not a laissez-faire reaction but a conscious choice on the part of the manager to let the people take the idea a little further by themselves.

This kind of manager uses the power of his or her position to teach the appropriate methods to subordinates. The effective managers use their knowledge to instruct without forcing their ideas on people.

The effective managers use the democratic part of their style to encourage creative exchanges about problem areas in the process. People in this environment are comfortable with being wrong. New hypotheses are tested, the results of which are discussed, and the next steps are planned. Consensus decision making is utilized when it is appropriate, but the group is guided to the right decision if it starts to focus on the wrong activities or ideas.

The effective manager uses the authoritarian part of his or her style to keep people focused. The process is efficient because people do not waste energy and money on activities that do not add value to the product. The manager points out

YOU IN THE PROCESS

the "red herrings" in the never-ending quest to improve the process.

One of the most difficult skills to acquire and use effectively is knowing how to step into a democratic type of discussion with a helpful but tactful directional comment. Several phrases can be utilized in this type of situation: Have you considered...? Have you looked at it this way? Let me play the "devil's advocate" with this idea.

It is difficult to be this type of manager. You will undoubtedly make mistakes in applying the different styles to the different process decisions. It takes effort and experience to improve your style. You have to work on improving all of your skills and be aware of how they affect your style. Certain situations requiring a particular skill may elicit a style from you that is inappropriate for that situation. Your response may hinder the development of the best solution.

For example, suppose one of your strengths is analysis. When a problem arises, you go into the process and immediately offer the solution with specific instructions. In this situation, the process improved immediately, but did the skills of the people in the process improve? How did this response improve their analytical skills? A better solution would have been to lead a discussion among the operators that would "develop" the solution you already see.

It doesn't always have to be your idea. Let someone else experience your success. It can be an excellent developmental tool.

If you make a wrong decision, admit it. Don't try to cover up your mistake or you will lose more respect more quickly from your people. Conversely, you will gain a considerable amount of respect if people see that you are human and willing to admit it.

How would you characterize your style? How about the style of your boss? How about the style of the supervisors working with or for you? How do the people in the process respond to the different managers? Which managers do they talk to when there is an issue to resolve in the process environment?

Take the time to jot down some thoughts on a piece of paper to compare the styles of these people. Once you

PART THREE: KNOWING YOURSELF

understand their styles, you can begin to develop more effective ways of dealing with them. For example, it would not be productive to approach someone with an authoritarian style head on. Alternatively, it would not make a democratic manager comfortable if you always responded to him or her in an authoritarian manner.

You should develop this same level of understanding for the people in the process. Recognizing and responding to a line operator's style will help you make his or her contributions to the process more efficient. It will also make him or her happier with the job because his or her individuality is being recognized.

If you are not sure of what style you should adopt, you can be sure of what style you should not adopt. Laissez-faire never works in a process environment. The purely authoritarian and purely democratic styles do not work 100 percent of the time. So, your goal is to avoid the extremes. If your style is already at one of the extremes, change it. It is never too late to improve yourself, and by improving yourself you improve the process.

CULTURE

You are a product of your environment. Charles Darwin demonstrated many years ago that creatures adapt to their surroundings. People and animals learn to survive based on how they expect the environment to affect them. The people working in a process environment react in the same way to the culture within the facility. They learn to survive and adapt whether it is a great or not-so-great environment in which to work.

You are rarely given control of a new facility. New facilities are not often constructed. If you are placed into a new process environment, you have a unique opportunity to create the culture within those walls. Even if you are constrained by your corporate culture, the freedom to selectively hire a new work force of employees and the experience of bringing the new group "down the process knowledge road" together can be very powerful culture-creating factors.

YOU IN THE PROCESS

In most situations, you will be assigned to an existing culture. While it may seem unique to you, it has cultural characteristics common to almost all processing facilities. The challenge for you is to determine where on the continuum the process is in relation to these characteristics. Here are some of the characteristics to look for: Are the people actively involved in trying to understand their process? Are the managers involved from the top down in the process? Do the managers HAVE it? Is there a "we can do anything" attitude toward resolving problems? Are the horizontal and vertical exchanges of information in the organization honest and open? Are people encouraged to be free and creative? Can "conventional wisdom" about the process be challenged? Are people happy? Do they feel informed, involved, and responsible? Do the people care about the process? Is quality defined and valued? Is the process achieving the desired efficiency and quality goals?

These characteristics can be used collectively to define a culture on a single continuum. This continuum is based on risk. Does the culture avoid or seek risk? At the extreme in a risk-averse culture, people are not encouraged to be creative, problem-solving explorers. People are not actively involved in trying to improve their understanding of the process each day. Conventional wisdom is not challenged, and managers work to maintain the status quo.

The risk-averse cultures and their processes eventually stagnate and lose their competitiveness in the marketplace. They may be efficient in their own eyes according to their outdated standards, but they have lost touch with the need, the drive, and the enthusiasm to develop more efficient process methods. These companies are often lulled into this position after having achieved an "untouchable" position in the industry or segment of the market.

At its extreme, a risk-seeking culture is not a desirable position for a company either. Small companies trying to define and find their niche in the market fall into this category. Investments in the best process equipment and systems may be scaled back as the risk of survival is compared with the risk of the rate of return.

PART THREE: KNOWING YOURSELF

People in this environment have to be free and creative and very involved in the process if it is going to survive. Conventional wisdom may change each week as more of the science of the process is discovered and understood. Specific quality goals may change as the effects of the different process parameters become more defined.

The best place for a culture to be is between the extremes, regardless of the size of the company. A well-managed process has a culture with just the right blend of characteristics from each extreme. The most successful cultures will be close to the center of the continuum but slightly on the risk-seeking side of center. They have to be there to create, improve, change, and perfect the process. Living close to the controlled edge creates the enthusiasm that spawns innovation and success.

Can one person make a difference? Yes, one person can make a difference, but remember the very important MPPY culture rule: *Processes and cultures always go downhill faster than they go uphill.*

Why? Because it requires a great amount of effort and commitment each day to keep improving the science of the process and the individuals working in it. When people are in an environment that does not support them or the process, they will not have the level of commitment required to keep the process "going uphill." It is easier to go through the motions of performing a job than it is to make every effort count.

You have to gain a thorough understanding of where the culture has been and where it is today before you can determine where it should go. Making uncalculated changes to the culture of a process can be disastrous. In a sociological context, cultures develop their trademark characteristics over many decades. You won't be around that long, so look for those aspects of the culture that can be changed and can have the most positive effect on the process. You may never be able to totally change it, but you may not have to change all of the culture to obtain excellent results.

YOU IN THE PROCESS

MANAGING CHANGE

If you have just been assigned to manage a process environment or to direct a project, then you are and will be perceived as a change agent by the people currently working in the process. It can be difficult to see immediately what is right or wrong with a process when you enter at the top. Yet the paradox here is that being at the top offers you the greatest opportunity to change the process.

Being the process or project manager provides you with the power to redirect the activities of the process. The machines can be rebuilt, reconfigured, replaced, slowed down, or speeded up. Work procedures can be redefined, redirected, expanded, or contracted. The number of different combinations of equipment and people can be quite large. Their combined effect can yield significant change in the process.

In most situations, change should be gradual. It is wise to avoid risk in your decision making until you understand the process. If the situation is more desperate, then take some calculated chances to improve the process.

How does a new process manager effect change in the entire process group? You have to win the confidence of the group. They have to believe in your message, and they must believe that you are doing the right thing for the business.

To accomplish this, the manager must HAVE it! Spend extra time in the process asking questions. How does this piece of equipment work? Is this good product? Why? What needs to be fixed? Why? Listen to the comments about what needs to be fixed and try to have it fixed quickly. All of your activities should demonstrate that you care about the process and want to improve it.

The phenomenon of asking the people about what is wrong and them telling you is at the heart of the consulting and therapeutic professions. The therapist can act as an impartial ear and judge. He or she can usually formulate several action plans from the analysis to address the real problem. If you are new to the process, then you can respond in the same impartial manner. If you are able to help the line operators improve the process, then the "therapeutic" momentum

PART THREE: KNOWING YOURSELF

will stay with you as you and your involved style are incorporated into the team.

In other situations, the change agent may come to you and your process without your specific approval. It may be a group of consultants brought in to help you identify and correct the problem areas. It may also be the case that the company is trying to change the operations strategy and wants to push the new value system out into the processing units as quickly as possible.

You must recognize that in either situation the change agents can be used to advance your knowledge of the process. Stay close to their activities. Walk around the process with them, ask them for regular updates on their findings, and take immediate action in response to their findings.

Although consultants may have some experience with your type of process, they still won't know as much about your facility as you do. Remember this if you start to feel defensive.

Position yourself between the change agent and the process in the implementation of the program. Gain control by having the other managers and line operators involved in the action plan. It is important to give the process people the control and responsibility for the action plan if the changes are going to have positive short- and long-term effects.

If you request help in the form of consultants or a specialist, make sure you know and they know what you expect from them. Again, stay close to their activities. Involve the best people from the process in the action plans. Make sure you get your money's worth and not a report that tells you what you told them. You can help avoid this common problem by defining the area they are examining in an open-ended manner. Don't bias their investigation. Remember, there are usually several ways to solve a problem.

A very common form of change in a process is the addition or subtraction of equipment from the process. It may be replacing a machine with a newer, faster, more efficient one or installing a new line.

In a small facility with a simple process, you may have to manage the process and the new project. In a larger facility with a more complicated process, someone else will and

YOU IN THE PROCESS

should manage the project. Your job is to keep the process on track and to position the organization for the changes that will occur when the new project becomes operational.

Changing the process by installing new equipment, replacing old equipment, or conducting a major overhaul of the existing equipment can only be successful if you understand the science of the process. How does the current line work? What are its critical process control points? Are these issues addressed in the new and different piece of equipment? How will the responsibilities of the supervisors and line operators change?

If you cannot answer these questions, you may be in for a very long startup curve for the equipment and a longer learning curve for its operators. You may not even be selecting the right equipment, and that is a very scary yet real possibility.

Whatever the change, and no matter how well you know your process, there will be something new for you to learn about your process during the change period. In your planning, you should anticipate the impact of the change on the existing process control points, but something new will surface and something will be missed. Expect it to happen and react to it professionally by providing the resources to fix it.

If you know your process and have planned the change carefully, you will have reduced the probability of a major problem. So, how do you plan? Start by having one person in charge of the project.

Identify all of the resources that are available to you and your project manager. Make sure to help your project manager through some tight spots by getting him or her extra resources if they are needed to complete the project successfully. Look for help from your corporate group. Maybe someone from another plant can be brought in to assist. Determine whether any specialized skills can be found in the area around the facility.

If the project involves the utility companies and/or the local government for zoning changes or permits, leave yourself plenty of time. Bureaucracies move slowly. It is best to contact each of these groups at the beginning of the planning stage if they are going to be involved.

The project manager should definitely involve the process people in planning the project. If the project manager is

PART THREE: KNOWING YOURSELF

from the process group at the facility, the communication patterns should be good, but don't assume that people talk just because everyone knows each other. Everyone may not know the same things.

Get line operators and supervisors in a room to discuss the change issues. Encourage people to be creative and to challenge the ideas in the project. Organize committees to review the specifics of each subprocess. A new project can be a great opportunity to spread the knowledge of the process more thoroughly to more people.

The project manager should publish minutes of all of the meetings. The agenda should include an update on existing items and a discussion of upcoming items. Always, always, *always* review the impact on the time line. Always, always, *always* review where the project costs are in comparison to the approved budget. A personal computer and a simple program can keep track of the purchase orders as they are issued.

If outside contractors are being used, have them submit daily or weekly summaries of the labor and material costs for the work performed. Do not allow more than 1 week between summaries, because it is too difficult to remember everything that was worked on. The contractors should present information on each specific project.

The project manager should make special notes on the impact the changes had on the original scope of the work. Change orders can affect the cost and timing of a project, and so they must be carefully monitored. Some contractors will use change orders to make up costs on a low-bid contract. Most important, a change order may unknowingly affect the intended science of the process. Make sure you analyze the effect on the process, the people, and the management of the project before changing any element of the plan.

On time and material projects, the cost control of the total project can quickly slip away from you if change orders are issued without an understanding of the current investment. Make sure there is an incentive to finish the project in a timely manner.

A PERT (Project Evaluation Responsibility Technique) or a CPM (Critical Path Method) chart can be very helpful in understanding the correct project flow. This type of planning and analytical tool can help you shorten the time line of the project by utilizing your resources more efficiently. Many books and articles on the subject have been written. Software is also available to help you make and change the charts quickly to play out "what if?" questions.

On small or large projects that will involve people from different organizations, spend the time prequalifying them. The time you invest up front getting to know the vendors or contractors will save you headaches later. Get comfortable with them in the beginning, because once the project starts you are going to be "comfortable" whether you like it or not.

INDEX

Accessibility, in communication, 64–65
Accidents, costs of, 83
Accounting, 43–48
   and receiving function, 106–107
Accounts payable function, 104
Agendas, 146
Assimilation, of new employee, 76–77
Authoritarian managers, 137, 138–139, 140

Back injuries, 84
Boss, communicating with, 130–135
Brainstorming sessions, 81

Change, managing, 143–147
Change orders, 146
Checklists, in developing quality standards, 17
Communication
   accessibility in, 64–65
   with boss, 130–135
   confidence in, 69
   enthusiasm in, 66
   with environment, 32
   honesty in, 63–64, 65, 69
   impact of unions on, 69–72
   with line supervisors, 91–92
   and managing change, 144, 146
   meetings, 79–82
   with process employees, 63–69
   visibility in, 65–66
Computers
   in report generation, 44, 45
   in monitoring environmental conditions, 38–39
Computer programs
   for combining inventory control and preventive maintenance, 38
   for computer meetings, 82
   for creating schedules, 102
   for electronic spreadsheets, 49–50
   for handling receivables, 107–108
   and statistics, 48–52
   for what–if analysis, 147
Consumers, and product development, 111
Continuous improvement, as a challenge, 4, 60, 121
   as a process, 5, 52, 55, 88, 120
   for the supervisor, 87
Contractors, use of outside, 146
Cost(s)
   controlling, 146
   efforts to reduce, 47
   identifying, 60
   fixed, 47
   pressure for lower and quality

PART THREE: KNOWING YOURSELF

control, 95–96
variable, 47
Cost-accounting standards, 46
Cost-quality relationship
   achieving the optimum, 30
   controlling, 16
CPM (Critical Path Method), 114, 7
Culture, of process facility, 140–142

Daily efficiency report, 45–46, 104
Democratic manager, 137, 138, 139, 140
Departmental process
   front office, 103–105
   line supervisor, 87–94
   product development, 109–115
   quality control department, 94–98
   receiving department, 106–109
   sanitation program, 99–103
   shipping department, 106–109
Discovery, 60–61

Early retirement programs, 134
Economic order quantity (EOQ) formula, 25
Electronic probes, 38–39
Employee involvement, in quality program, 61–62
Enthusiasm, in communication, 66
Environmental conditions, 38–41
Ethics, in process environment, 2

Facetime, 91
Feedback, providing positive, 90, 123
Flexible manufacturing systems, designing, 26–27
Frequency charts, 51
Frequency distribution of parts, 36
Front office, 103–105

Headcounts
   fluctuations in, 104
   reducing, 10
Heating, ventilation, and air conditioning process (HVAC), 39–41
Hiring process, 72–77, 92, 97
Honesty, in communication, 63–64, 65, 69

Insurance costs, 83
Interview process, in hiring, 74–75
Inventory
   problems in storing, 27–28
   procedures for taking, 47
   and shopping and receiving process, 107–108
   zeroing items out of, 46

Japanese, 29–30
   cultural characteristics, 29–30
   success of, 29
Job description, 73
Job qualifications, 75–76

Laboratory
   benefits of, in process environment, 61
   keeping quality people out of, 98
   prototyping of product improvement and cost reduction tests in, 110
   use of outside, 19
Laissez–faire managers, 137–138, 140
Leadership, styles of, 137–140
Line operators
   and quality control groups, 96–98
   and sanitation, 100
Line supervisors, 87–94
Loyalty, 133–135

INDEX

Machine operators, training of, to perform maintenance, 34
Maintenance. *See also* Preventive maintenance
    correlation with sanitation, 102
    and delayed repairs, 92–93
    and handling equipment breakdowns, 131–132
    and keeping work area clean, 35–36
    parts in, 35–38
    rotating of personnel in, 34
    training of personnel in, 33
Maintenance group, management of, 35
Management. See also Line supervisors; Process managers; Project managers
    of change, 143–147
    identifying strengths in, 129
    and level of knowledge, 65
    styles of, 137–140
Managing the Process, the People, and Yourself (MPPY) philosophy, 4
    operation rules in, 5, 15, 16, 18, 32, 33, 45, 63, 64, 69, 77, 89, 90, 95, 110, 111, 122, 128–129, 131, 142
Material consumption, variable costs for, 47
Material Requirements Planning (MRP) systems, 24
Material specifications, effects of, on quality standards, 19–22
Measurement, and accounting for the process, 43–48
Meetings, 79–82
    agendas for, 146
    and change management, 146
    quality review, 96–97
    safety, 83–84, 85
Mindset
    in quality control, 94–95
    in safety, 83–84
Moving, 134–135
Multiple shift operation, managing, 91–94

Names, use of first, 64
National Labor Relations Board, 70
Negotiation
    improving skills in, 23
    win–win concept in, 71–72
Nonunion setting, communication in, 70–71

Occupational Safety and Health Administration (OSHA), 70
Organization charts, 122
    in hiring process, 76

Parameters, changing, 18
Payroll, generating, 104
PERT (Project Evaluation Responsibility Technique, 146–147
Preventive maintenance, 31–35
Process
    accounting for, 43–48
    analyzing when working well, 18
    definition of, 1
    gathering information on, 21
    identifying valuable and non–valuable activities, 122–123
    importance of understanding entire, 9–11
    knowledge of, 16–17
    managing technical aspects of, 55
    success of, 13–17
Process employees, communication with, 63, 67–69
Process environment
    benefits of laboratory in, 61
    best managers in, 4

PART THREE: KNOWING YOURSELF

cleanliness and appearance of, 99–103
collecting data in, 45
communication in, 32, 63–69
conditions in, 38–41
ethics in, 2
examples of, 1–2
maintenance parts in, 35–38
preventive maintenance in, 31–35
quality control in, 94–98
range in sizes, 1
role of accountant in, 44–45
safety in, 2, 82–85
support functions in, 55
team concept in, 58
use of statistics in, 50–51
Process facility, culture of, 140–142
Process improvement
  achieving, 123–124
  managing and improving yourself, 124–130
Process manager
  best, 4
  commitment of, 59–60
  and data collection, 50
  goals of, 10–11, 122
  and quality control, 98
  responsibilities of, 121–135
  role of, 121
  skills needed for success, 4–5, 119–120
Process–quality relationship, 96–98
Product development, 109–115
  evaluating feasibility of new ideas, 112–113
  factors in determining success in, 15–16
  process in, 14–15
Product hold, 21–22
Product recall, 22
Production costs, 16–17

Project manager, and managing change, 145–146
Promotions, 76, 134–135
Purchase orders
  creating, 104
  signing, 46
Purchasing, 22–24
Purchasing agents
  relationship with suppliers, 23
  reliance on sales forecasts, 24–25, 27

Quality
  commitment to, 58
  definition of, 13–17
  role of people in, 57–63
  specifications of, 13
  standards of, 13
Quality control department, 94–98
Quality control groups, and line operators, 96–98
Quality program, employee involvement in, 61–62
Quality review discussions, involvement of line supervisors in, 91
Quality standards, 17–22
  making changes in, 46
  and materials specifications, 19–22

Receiving department, 106–109
Recognition programs, 68–69
Reports
  daily efficiency, 45–46, 104
  processing information for corporate, 105
  scheduling, 102–103
  test results in, 62–63
  value of, 43–44

Safety, in process environment, 2, 82–85
Safety committee meetings, 85

INDEX

Sales forecast, reliance of purchasing agent on, 24–25
Sampling process, 106
Sanitation program, 99–103
 developing plan for, 101–102
Scrap analysis, 47–48
Self–image, creating positive, 127
Shipping counts, accuracy of, 108
Shipping department, 106–109
Statistical process control, 18, 51
Statistics
 and computers, 48–52
 in process environments, 50–51
Stress, managing, 124–128

Teamwork, in multiple shift operations, 92–94
Telephone calls, handling, 103, 105
Termination process, 77–79
Time and material projects, 146
Time cards, initialing, 46

Training
 of employees in test methods, 62
 of machine operators in maintenance, 34
Training manuals, development of, 33, 85
Transfer
 of employees, 134–135
 of knowledge, 134

Union, impact of, on communication, 69–72

Visibility, in communication, 65–66

Wagner Act, 69–70
What–if problem solving, 105, 123, 147
Win–win concept in negotiations, 71–72
Workers' Compensation costs, 83

153

DISCARDED

JUN 30 2025

ASHEVILLE-BUNCOMBE
TECHNICAL COMMUNITY COLLEGE

3 3312 00039 5947

TS 155 .W457 1993

Werner, Joseph G., 1951-

Managing the process, the
people, and yourself